DAILY SPATIAL MOBILITIES

Transport and Mobility Series

Series Editors: Professor Brian Graham, Professor of Human Geography, University of Ulster, UK and Richard Knowles, Professor of Transport Geography, University of Salford, UK, on behalf of the Royal Geographical Society (with the Institute of British Geographers) Transport Geography Research Group (TGRG).

The inception of this series marks a major resurgence of geographical research into transport and mobility. Reflecting the dynamic relationships between socio-spatial behaviour and change, it acts as a forum for cutting-edge research into transport and mobility, and for innovative and decisive debates on the formulation and repercussions of transport policy making.

Also in the series

Territorial Implications of High Speed Rail
A Spanish Perspective
Edited by José M. de Ureña
ISBN 978 1 4094 3052 0

Sustainable Transport, Mobility Management and Travel Plans
Marcus Enoch
ISBN 978 0 7546 7939 4

Transition towards Sustainable Mobility
The Role of Instruments, Individuals and Institutions
Edited by Harry Geerlings, Yoram Shiftan and Dominic Stead
ISBN 978 1 4094 2469 7

Integrating Seaports and Trade Corridors
Edited by Peter Hall, Robert J. McCalla, Claude Comtois and Brian Slack
ISBN 978 1 4094 0400 2

International Business Travel in the Global Economy
*Edited by Jonathan V. Beaverstock, Ben Derudder, James Faulconbridge
and Frank Witlox*
ISBN 978 0 7546 7942 4

Daily Spatial Mobilities
Physical and Virtual

AHARON KELLERMAN
University of Haifa, Israel and Zefat Academic College, Israel

Routledge
Taylor & Francis Group

LONDON AND NEW YORK

First published 2012 by Ashgate Publishing

2 Park Square, Milton Park, Abingdon, Oxon OX14 4RN
711 Third Avenue, New York, NY 10017, USA

Routledge is an imprint of the Taylor & Francis Group, an informa business

First issued in paperback 2016

British Library Cataloguing in Publication Data
Kellerman, Aharon.
 Daily spatial mobilities : physical and virtual. --
 (Transport and mobility series)
 1. Spatial behavior. 2. Information technology--Social
 aspects. 3. Mobile communication systems--Social aspects.
 4. Transportation--Social aspects.
 I. Title II. Series
 304.2'3-dc23

Library of Congress Cataloging-in-Publication Data
Kellerman, Aharon.
 Daily spatial mobilities : physical and virtual / by Aharon Kellerman.
   p. cm. -- (Transport and mobility)
 Includes bibliographical references and index.
 ISBN 978-1-4094-2362-1 (hardback)
 1. Space in economics. 2. Social mobility. 3. Residential mobility. I. Title.

 HT388.K454 2012
 304.8--dc23

2011053329

 ISBN 978-1-4094-2362-1 (hbk)
 ISBN 978-1-138-26856-2 (pbk)

Contents

PART III: SPATIAL IMPLICATIONS

List of Figures and Tables

Figures

Tables

Preface

This book attempts to explore a facet of our lives that we often take for granted, namely our daily spatial mobilities, whether it is commuting to work or e-mail exchanges. The wide spectrum of daily spatial mobilities, including both corporeal and virtual mobilities, requires, I believe, some systematic treatment looking for their roots, expressions, potential, media, meanings and impacts, and this is what I tried to develop and offer in this book.

In some ways this book constitutes a continuation of my previous book entitled *Personal Mobilities* (2006a), which at the time continued my *The Internet on Earth: A Geography of Information* (2002). *Personal Mobilities* attempted to present the case of personal mobilities, both corporeal and virtual, defined as "moving of the self by the self" (p. 1), such as in driving and in the use of the Internet, whereas *The Internet on Earth* proposed a systematic geography for information focusing on the Internet. This book treats, like *Personal Mobilities*, both physical and virtual mobilities, but this time within the wider context of routine daily spatial mobilities. As such it discusses not only personal mobilities but public transportation as well, as part-and-parcel of daily spatial mobilities at large. This volume might be considered a first attempt to treat daily, routine spatial mobilities as such, and thus distinguish them from the non-routine mobilities, notably tourism and migration.

Following my interest in mobilities at large, and in personal mobilities in particular, I have focused my research since 2006 on a variety of aspects which have formed the basis for most of the following chapters. When reflecting, some time ago, on the common thread among all these mobility aspects, I noticed that the common denominator was routine mobilities, bringing me to the need to examine this large and varied family of mobilities in a more systematic way, and hence produce this book.

In the writing of some of the chapters I was able to enjoy the criticism and advice of Sven Kesselring and Maria Paradiso who read them as draft articles. Obviously, responsibility for the writings is mine alone. The Editors for the Ashgate *Transport and Mobility* series showed a keen interest in the topic and the book idea and provided valued comments and support for its advancement. The Research Authority of the University of Haifa assisted the completion of this book through the funding of index preparation.

Special thanks go to my wife, Michal, and to my entire family, for their bearing with me with great patience and understanding at intensive and sometimes unusual times and places of study and writing. The book is dedicated to my grandchildren Itai-Baruch, Shaked, and Nadav-Yehuda, who were born during the various phases of its development and writing.

Aharon Kellerman
November 2011

Acknowledgements

The following previous publications of the author were used for the writing of this book: *Personal Mobilities*. 2006. London and New York: Routledge; Cyberspace classification and cognition: Information and communications cyberspaces. *Journal of Urban Technology*, 14, 2007: 5-32; Geographical location in the information age: From destiny to opportunity? *GeoJournal*, 70, 2007: 195-211 (with M. Paradiso); International airports: Passengers in an environment of 'authorities'. *Mobilities*, 3, 2008: 161-78; End of spatial reorganization?: Urban landscapes of personal mobilities in the information age. *Journal of Urban Technology*, 16, 2009: 47-61; Mobile broadband services and the availability of instant access to cyberspace. *Environment and Planning A*, 42, 2010: 2990-3005; *Mobility* or *mobilities*: Terrestrial, virtual and aerial categories or entities? *Journal of Transport Geography*, 19, 2011: 729-37; Potential mobilities. *Mobilities*, 7, 2012: 171-183; Business travel and leisure tourism: Comparative trends in a globalizing world, in *International Business Travel in the Global Economy*, edited by J.V. Beaverstock, B. Derudder, J. Faulconbridge and F. Wiltox. Farnham: Ashgate, 2010: 165-75.

Dedicated to Itai-Baruch, Shaked, and Nadav-Yehuda

Chapter 1

Introduction

This chapter will begin with a brief presentation of mobility at large as a contemporary field of study. It will then explore the two major branches of mobility studies, social and spatial, and the interrelationships between them. The chapter will then move to a focus on the specific class of daily spatial mobilities as compared to non-daily mobilities. Daily mobilities include, among others, mobilities for commuting, shopping, social ties, information, banking, news, studies, business meetings, etc. These mobilities are typified by their being two-way mobilities, frequently performed and constituting a major element of our daily, routine lives, inclusive of both corporeal and/or virtual mobilities. Non-daily mobilities, on the other hand, include the two-way mobility of pleasure tourism and the one-way mobilities of residential change and migration which normally involve social change. Next in this chapter will be elaborations on key concepts for the study of daily spatial mobilities. The chapter will conclude with a brief presentation of the chapters that follow.

Mobility

Human mobility in its most general and basic sense may be referred to as shifting, or the human ability to shift. Such shifts may refer, first, to the ability of the human body to move across space, or to the ability of humans to move their limbs. It may further relate to the ability of humans to move themselves using either ancient or contemporary mobility technologies. The shifting of humans over space always involves displacement, whether minor and repetitive as in daily commuting, or whether major and one-way one as in migration. These shifting abilities, which are mainly spatial, or 'horizontal' in nature, have been extended to various other senses, including social, or 'vertical', mobility, namely the shifting of people from one social level or occupation to another (see *Oxford English Dictionary* 2010, Cresswell 2006a: 20). Social shifting or mobility always implies change, whether in social position or in social status, as compared to displacement which constitutes the essence of spatial mobility, in which change may frequently emerge as cumulative, through numerous movements. Mobility is, thus, a multifaceted term. For human mind and action it was described as including:

> From displacement from one location to another to the freedom of movement which is symbolically equated with social mobility, to the feelings of pleasure in effortless flight which has roots in infancy, to the fundamental psychic link

of motion with causality and subjecthood first described by Aristotle. But mobility also suggests the opposite of subjecthood, the freely displaceable and substitutable part, machine or human, which enables mass production and a consequent standardization brought to the social as well as economic realm (Morse 1998: 112; see also Buliung 2011).

The study of mobility is not just interested in the shifts themselves, but no less in their contexts and significances. As such it goes beyond the traditional study of transportation geography (Shaw and Hesse 2010; Bissell *et al.* 2011). Any shift or mobility "is *given* or inscribed with meaning. Furthermore, the way it is given meaning is dependent upon the context in which it occurs and who decides upon the significance it is given" (Adey 2010: 36).

Social mobility

As we just mentioned, the term mobility has received a sociological connotation within the context of *social mobility*, referring to status transitions of individuals and social groups along societal strata. The study of social mobility is beyond the scope of this volume, but one may potentially argue that social and spatial mobilities are interrelated, in the sense that upward social mobility may imply extended and increased spatial mobility *vis-à-vis* an enhanced ability to purchase and use automobiles and telecommunications services. Also, one could assume an extended ability to use and benefit from these mobility technologies, notably the Internet, if elevated social status is accompanied by additional education. Such a relationship may potentially also go the other way around: increased physical and/ or virtual spatial mobility may imply wider information and physical reach, thus providing stimulation and opportunity for social mobility. However, the mutual relationships between spatial and social mobilities are complex (Urry 2007: 8, Adey 2010: 37-8), bringing some to suggest that there is no connection anymore between the two mobility forms (Kaufmann 2002: 12-3, Bonss and Kesselring 2004), at least as far as physical mobility is concerned. For virtual mobility, via telecommunications, however, it was argued that "it is no longer geographical space that differentiates but virtual space", and "the more telecommunications there is, the more social mobility" (Kaufmann 2002: 29).

Spatial mobility

The mobility of human beings, in the sense of humans moving over 'horizontal' space, rather than the more veteran notion of 'vertical' social mobility, has received growing attention in recent years, frequently blurring a clear distinction between 'spatial mobility' and 'mobility' at large. Spatial mobility has been viewed as a positive societal trend and force and as an integral part of the second modernity, involving wide social implications (see e.g. Urry 2000, 2007). Spatial mobility was variously defined as an activity or social dimension: "geographical displacement, i.e. the movement of entities from an origin to a destination along a specific trajectory that can be described in terms of space and time" (Kaufmann *et al.* 2004: 746). "Spatial mobility is not an interstice, or a neutral liaison time between a point of origin and a destination. It is a structuring dimension of social life and of social integration" (Kaufmann 2002: 103, see also Urry 2000, 2007). Also, *"mobility is polysemic and does not itself reveal what underlies it"* (Kaufmann 2002: 101-3). Thus, for Baudrillard (1966: 66) "effortless mobility entails a kind of pleasure that is unrealistic, a kind of suspension of existence, a kind of absence of responsibility".

A basic definition for spatial mobility, from the perspective of transportation geography, views mobility as ability: *"Mobility* refers to the ability to move between different activity sites" (Hanson 1995: 4). By the same token, *movement* was described as "the idea of an act of displacement that allows objects, people, ideas – things – to get between locations" (Cresswell 2001a: 14). It is interesting to note that movement, or the mobile, was defined here through displacement which is a negative form of the term place, a term which traditionally describes the fixed or the sedentary rather than the mobile! This physical connotation for movement/mobility is typical within human geography, referring to the very human ability to move oneself in the sense of daily physical spatial mobility (see e.g. Ogden 2000, Urry 2004a). Others, notably sociologists, preferred to refer to mobility over space as *spatial mobility* (Kaufmann 2002). Bonss and Kesselring (2004: 5), on the other hand, provided a rather social and more restricted definition for mobility: "an actor's competence to realize certain projects and plans while being 'on the move'".

Spatial mobility, stemming from 'push and pull' motivations, which we will discuss in the next chapter, constitutes foremost a constant, omnipresent "displacement of something across, over and through space" (Adey 2010: 13, see also Cresswell 2006a: 1-2, Morse 1998: 112). From the perspective of *homo viator* [mobile person] (Eyerman and Löfgren 1995), this displacement is practiced, experienced and embodied (Cresswell 2006a: 3). As such, spatial mobility is a meaningful condition, implying progress, freedom, opportunity and modernity (Cresswell 2006a: 1-2), as well as speed (Prato and Trivero 1985: 40, Virilio 1983: 45) and extensibility (Adams 1995, Kwan 2001).

The recent telecommunications/information revolution has loaded the term *mobility* with yet another meaning, namely the human ability to make a rather abstract entity, information, flow electronically. Such electronically-transmitted information may constitute a virtual extension of the self, through a phone call or an e-mail, or it may constitute more public pieces of information available through websites, and thus not transmitted as one-to-one or one-to-specific several receivers by an end-user. The mobility of information constitutes *virtual spatial mobility*. The mobility of information may be viewed as mobility for itself, or it may be defined in light of physical mobility. *"Virtual mobility* refers to the substitution of electronic transfers and exchanges for physical transport activities"

(Janelle 2004: 86). Urry (1999) named the virtual information flows through the Internet *weightless traveling*, whereas *imaginative traveling* refers to such flows through television broadcasting. Though television broadcasts amount to one-way public transmission of predetermined information, they were compared to personal physical mobility via automobiles by Bachmair (1991) who claimed that "television succeeded because it broadened and extended lifestyles associated with the motor-car; primarily those concerned with *mobility* as a shaping principle of communication" (522). Others named mobility *vis-à-vis* television, as *transport of the mind*: "Television turns out to be related to the motor car and the aeroplane as a means of transport of the mind" (Rudolf Arnheim, quoted in Morse 1998: 99).

Displacement is possible for three sorts of movables: people, objects and information/knowledge (Urry 2007: 7-8, 47, Kaufmann 2002), and these three movables may be differentiated by their mobility flexibility using a state of matter metaphor (Kellerman 1993: 160). Moving information is as flexible as gas, easily changing modes, shapes and volume, and its transmission being instant. People's corporeal mobility is like liquids, in people's ability to change travel modes, and in their ability to be partially self-motored, though mobility usually requires some preparations. Moving objects is the moving of solids and is thus slower, always requiring handling for the very mobility of objects. All three movables are human in some way, since objects and information are sent by people and for people, sometimes replacing human corporeal mobility. The mobilities of people and information have become integrated, as communications permit the coordination and management of physical mobility. Furthermore, it has become possible for individuals to move corporeally while communicating virtually. Still, the mobilities of people and objects are also interrelated: "There are objects that enable people to travel across distance; there are objects that enable people to travel forming complex hybrids...there are objects and people that move together" (Urry 2007: 50). The mass moving of objects has become increasingly organized and controllable through logistics and modal transportation, side by side with the opposite trend for the daily mobilities of humans who prefer to move individually and, thus growingly possess personal mobility technologies.

As we have noted through its various definitions, *spatial mobility*, physical as well as virtual, constitutes a double phenomenon. On the one hand, it relates to the *ability* to cross certain distances within certain time units. By human nature this ability is performed physically through walking or running, and virtually by speaking or shouting. However, in its more contemporary context, this ability may be measured through access to, and availability of, transportation and communications means. Personal spatial mobility, as compared to public spatial mobility, may be measured by the rate of *adoption* of transportation and communications means by households. Side by side with spatial mobility constituting a human ability, spatial mobility also relates to the very *use* of technology-based mobility media, or the performed *movements* by actors. Three possible relations may potentially develop between physical and virtual personal mobility media, when virtual mobility media become available next to physical mobility: substitution (i.e. virtual mobility replacing any physical movements, such as performing banking actions through the Web instead of at the bank branch), complementarity (i.e. physical mobility is complemented by virtual ones, such as phone calls preparing for any physical movements), or additivity (a new movement is added through virtual mobility, such as information search through *Google*, or the use of mobile phone while driving or riding a car) (see Kellerman 2006a). In line with the fluidity metaphor, Urry (2000: 32; see also Shields 1997) pointed to the possible distinctions among mobilities by their rates of flow, their viscosity, depth, consistency, and degree of confinement.

Daily spatial mobilities

As mentioned already, spatial mobility includes both routine cyclical rides and walks, as well as long-distance (in space) and longer range (in time) human movements of migration, tourism (or travel), residential change, mobile resistance movements, and the wandering of youngsters, etc. (see e.g. Verstraete and Cresswell 2002, Kaufmann 2002: 35, 2004, Urry 2000: 145-7). Some of these longer range and non-daily mobilities might be two-way, notably tourism, whereas others are one-way, notably migrations. In a slightly different way, spatial mobilities may be divided into reversible (daily trips and travel), and irreversible (migration and residential change) (Kaufmann 2002: 24-6). Thus, Kaufmann (2002: 40) sees spatial mobility in a seemingly wider sense of purpose, consisting not only of travel and daily mobility, but of migration and residential mobility, as well. His view of spatial mobility does not include virtual mobility in it (as stated explicitly on p. 46, but see also p. 35), but he treats jointly one-way or irreversible mobilities, such as migration and residential mobilities, with two-way, reversible movements, such as daily movements and travel. To some degree, the analyses of migrations and mobility in films advanced by Cresswell (2001b, 2002) follow a similar line. Spatial mobility in the form of migration has been shown by Cresswell (2006a) to constitute a major cultural experience.

All four categories of spatial corporeal mobility (daily, travel, migration and residential change) were viewed as sharing some similarities:

> None of the four areas of spatial mobility analysis has entirely done away with the dual definition of mobility. However, they generally make *social change* (understood as change in social status or role) correspond to *movement*: we move about on a daily basis in order to change roles; we travel to confront otherness and escape our daily grind; we move house when our lives are touched by change. With the development of rapid transport and telecommunications networks in the 1990s, the parallel between movement and change generally dissipated (Kaufmann 2009: 49).

The basic assumption for this proposed volume is that human spatial mobilities may be divided into two major types. One, which this book attempts to address, includes daily mobilities, such as for commuting, shopping, social ties, information, banking, news, studies, business meetings, etc. These mobilities are typified by their being two-way; frequently performed; constituting a major element of our daily routine lives; and inclusive of both corporeal and/or virtual mobilities. The other type of human spatial mobilities consists of non-daily mobilities which include both two-way and one-way mobilities, some of which reflecting structural-social change. This second type of mobilities includes the two-way mobility of pleasure tourism and the one-way mobilities of domestic residential change and international immigration, both of which involving social change. Following McKenzie (1927), Kaufmann (2009) suggested a distinction between daily movements which he termed *fluidity* and non-daily movements which he called *mobility*. Fluidity is supposed to have no effect on one's life as compared to non-daily mobilities. As we will see in the next chapter the roots for this differentiation might be deeper, but repeated daily movements have strong impact on one's life as a mobile person, in terms of a person's extent of activity space, accessibility and extensibility, potentially bringing about non-daily movements of residential change, travel, and even migration.

Much of our daily spatial mobility is fully virtual, including telephone calls, e-mails and Internet surfing, whereas non-daily mobilities are foremost corporeal, though aided by virtual ones, for instance through hotel reservations for a vacation made over the Internet. The motivations and triggers for daily mobilities are different from those of non-daily ones, and so are the preparations for movements, which for non-daily mobilities may be much more extensive. Behavior while on the move might again be different for daily mobilities than for non-daily ones, when being, for example, on vacation and travelling longer hours. In general, Bán (2007: 289) suggested that "with all of his or her everyday movement, *Homo sapiens* transformed into *Homo mobilis*".

Daily spatial mobilities, like any other human mobilities and activities at large, may be differentiated along major demographic dividing lines such as gender (see e.g. Blumen and Kellerman 1990) and age (see e.g. Burnett and Lucas 2010), and obviously along income. However, these differentiations will not be explored in this volume as they require separate and rather distinct conceptual settings.

Key terms for daily spatial mobilities

Before proceeding to the following chapters we will get briefly acquainted with some twelve key terms for the study of daily spatial mobilities. Their order of presentation follows three spheres: context-environment; movement; and spatial extent:

Context-environment: Information society; globalization; space of flows; networking; fixity.

Movement: Directionality; circularity; speed.

Spatial extent: Extensibility; accessibility; time-space compression; distanciation.

We will note first some introductory remarks on the three spheres containing the eleven key terms for daily spatial mobilities, prior to a separate treatment of each of them. The first of these three spheres is the context or the environment of daily spatial mobilities. The context for daily mobility may refer basically to mobility at large, not just to the daily one. However, it is of much significance for the locations of origins for daily movements, since the context of mobility may determine the media used for these and the moving of information (information society); the geographical options for mobility (through globalization), the system of flows (through the space of flows), the framework for much of them (through networking), and the role of places of origin (through fixity). These dimensions seem at first sight as relating mainly to virtual mobility, but they are of much importance to daily physical movements as well. We will attempt to discover this later on. As for the movements themselves, it is important to note their directions (directionality), their relations with the points of origin (circularity), as well as their speed. Finally, the spatial extent of movements and of mobility at large has much to do with destinations, potential ones as well as actual ones. Thus, we will explore the potential for reaching out (extensibility), the potential for routing as well as access to destinations (accessibility), the nature of reaching any destinations located in a globally reachable world (time-space compression), and finally the spatial expansion of target destinations and its significance (distanciation). We will treat jointly corporeal and virtual daily mobilities, because, as we will see later, they may often tend to converge.

Information society

The information society is the context within which the development and massive use of ICTs (information and communications technologies) has emerged. The information society is, further, the context within which the production, transmission and consumption of information has flourished. Both dimensions, ICTS and information, will accompany us when exploring daily spatial mobilities throughout this volume. Castells (2000: 21, n. 31) differentiated between 'information society' and 'informational society'. Whereas he related 'information society' to the role of information in society, which has always been of some importance, 'informational society' was attributed to "a specific form of social organization in which information generation, processing, and transmission become the fundamental sources of productivity and power because of new technological conditions". It so happened that what Castells called informational society is what has come to be generally termed information society. The emergence of the information society since the 1960s may be traced along three phases, which have been elaborated elsewhere: information-rich; information-based; and information-dominated (Kellerman 2000).

The definition for the information society favored by Webster (2006: 9) states that: "*theoretical knowledge/information* is at the core of how we conduct ourselves these days". Such a definition related more to a process than to a state of society, and it included, in Webster's analysis, several dimensions: technological, economic, occupational, spatial, and cultural (Webster 2006). For our focus here on daily spatial mobilities, the spatial dimension is of some interest. The information society is based on cyberspace, which is the virtual entity hosting the Internet, and which will be discussed in detail in Chapter 7. However, we may define information society already now from a spatial perspective as a society in which information is widely available electronically anywhere, and a society in which interpersonal electronic communications is possible from anywhere to everywhere in a variety of forms. The spatial organization of the information society is complex behind the scenes, in terms of the location of, and interactions among, the computer servers which make it function. However, from the perspectives of individual information users, continuous technological developments have made it increasingly simple to manipulate information and to operate information devices. More importantly, within contemporary information society, each human being may be considered as a node in the global information/communications system when carrying a mobile phone, notably smartphones, which permit access to the Internet and other information sources.

Globalization

One could view the essence of globalization as transitions in the location and nature of production, distribution and consumption of products, services, capital and information, from locally and domestically based systems to highly integrated global ones (Dicken 2007: 8-9). From a social perspective, "globalization involves the intensification of social exchanges of various sorts on a global scale" (Drori 2007: 303). To a large degree, the massive adoption of new communications technologies (notably the Internet and mobile devices), as well as the upgrading of older ones (notably through digital telephony and the provision of telephone and Internet services via cable TV and Asymmetric Digital Subscriber Line (ADSL)) have implied direct globalization of virtual mobility and contacts. By 'direct globalization' we mean the very geographical expansion of destinations for daily contacts with other countries. This direct globalization applies to international telephone calls, the tariffs for which have been drastically reduced or even provided free of charge over the Internet through VoIP (Voice over Internet Protocol) services. 'Indirect globalization' may refer to daily local and out-of-town physical mobility, as well as to domestic communications, all stemming from the globalization of activities, for example daily commuting to work for multinational corporations, or working for domestic companies exporting globally, or, on the consumption side, having some food or enjoying entertainment in a facility owned or operated by global chains. Generally, Urry (2000: 32) argued that "behaviour and motivation are less societally produced and reproduced but are the effect of a more globally organised culture that increasingly breaks free from each and every society".

Three dimensions of globalization are of special significance for daily spatial mobilities: fluids; networks; and scapes. Urry (2000, 2003a) used the metaphor of fluids as a general term for abstract entities which are being moved globally, such as information, capital, risks, etc., and "any such fluid can be distinguished in terms of the *rate* of flow, its *viscosity*, the *depth*, its *consistency*, and its degree of *confinement*" (Urry 2000: 32). The flows of these fluids are channeled through networks with varying degrees of flexibility. Social networks are most flexible ones, with participants joining and leaving them freely, preferring to keep their anonymity in many cases. Networking can further be commercial, with identities revealed and used by network managements. Intranets are more rigid networks, open only to employees of specific companies. Even stricter in their access, and for obvious reasons, are banking or inter-bank networks, such as SWIFT (Society for Worldwide Interbank Financial Telecommunication). Castells (2001) viewed the Internet as a medium which facilitates the establishment and functioning of networks for business management. These networks, on their part, constitute flexible, adaptable, and coordinating organizational tools, which permit a *many to many* communications mode on a global scale. He further viewed Internet networking as reflecting globalization, freedom, and telecommunications technology.

Appadurai (1990) distinguished five dimensions of global cultural flows, which he termed *scapes*, since, similarly to landscapes, they are looked upon differently by various actors, and since they are fluid and irregular. These five scapes are: *ethnoscapes* (the migration of workers); *mediascapes* (television, movies, magazines, etc.); *technoscapes* (technology); *finanscapes* (finances); and *ideoscapes* (ideologies). To this list one can add *commodiscapes*, for commodities which carry some cultural identity or messages on a global scale, such as pizza as an Italian food, or Japanese cars (see also Knox 1995: 245), and *infoscapes*, for information and knowledge, globally transmitted mainly via the Internet, and thus not included in mediascapes (Kellerman 2002).

Space of flows

Globalization was seen by Castells (1989) as leading to a lower significance of local places, coupled with the emergence of a global *space of flows*. The space of flows was defined as "the material organization of time-sharing social practices that work through flows" (Castells 2000: 442), in which 'flows' include all possible ones, except for people: capital, information, technology, organizational interaction, images, sounds, and symbols. The space of flows, as developed by Castells (2000: 448), consists of three layers: the first one is technology, constituting a *circuit of electronic exchanges* embodied in networked cities; the second is a layer of places, *nodes and hubs*, hierarchically organized and topped by global cities, which serve as major loci of information production. The third layer is people, the *managerial elites*, charged with the directional functions for the space of flows. However, the space of flows is the space in which many people, not only the managerial elite, operate daily, mainly virtually through communications channels, whether exchanging e-mails or browsing the Web. For the managerial elite, the space of flows implies also physical mobility through international, mostly aerial, travel.

Networking

The standardization of electronic transmissions of signals of all kinds (textual, visual, and audio) has permitted the emergence of electronic networking, mainly through the Internet at large, and even more so through Web 2.0 applications such as *Facebook*, side by side with traditional social networks, based on face-to-face meetings, commercial impersonal networks such as TV broadcasts, as well as public physical networks, such as road systems. The networking dimension of the Internet has been accentuated in the 2000s with the emergence of Web 2.0, focusing on social networking through users' blogs, *Facebook*, *Twitter*, etc. Despite the seeming autonomous nature of personal virtual mobilities, communications may undergo surveillance and turned into recorded mobility. Internet use by surfers may be recorded by commercial companies in order to channel proper advertisements, mobile telephone users may be located through GPS (Globally Positioned Satellites) and other technologies, and telephone calls may be illegally recorded as well. Though these surveillance activities present diversified motivations they imply an almost total potential and/or actual surveillance of personal mobilities.

Castells (2000) developed the notion of *network society*. "In all sectors of society we are witnessing a transformation in how their constitutive processes are organized, a shift from hierarchies to networks. This transformation is as much an organizational as a cultural question" (Stalder 2006: 1). The new network society refers to a global entity within which Castells (2000) proposed the emergence of the 'space of flows' which we just discussed. Wellman (2001a) viewed wireless communications as expressing a new phase in social communications and networkings. He termed non-technological communications as *door-to-door* communications, relevant to social relations within traditional, physical-place bounded communities. The automobile and the telephone have permitted the development of a second phase of social relations and networking, *place-to-place* ones, replacing some of the local door-to-door relations, and permitting people to easily communicate with other places. The Internet has enhanced place-to-place networks through its provision of continuous communications. Placeless wireless communications have implied the emergence of a third phase, *person-to-person* communications, detached from household and office locations and their communications infrastructure. Our daily movements, as to where we go and with whom we communicate, reflect networks of peers and friends, mostly reached through electronic communications media.

"The contradictory experience of being somewhere and nowhere at the same time is perhaps the most obvious cognitive dissonance resulting from the use of the WWW" (Kwan 2001: 26). The emergence of global social networks and the growing interaction with global information networks may bear upon the sense of place of users, and on processes of place production, notably since Internet users are co-present in physical and virtual places simultaneously. Network society may further be interpreted in the light of Lefebvre's thesis on *The Production of Space* (1991). Lefebvre insisted that space is not a thing but rather a social process unfolding in everyday life, and thus one may ask: "Are we engaged in the production of new spaces and new social relations, or merely simulating social structures in a hyperreal form? How does our experience of the global and the local, the public and the private, alter in a network society?" (Nunes 2006: xxiii).

Fixity

Mobility at large, and daily spatial mobility in particular, cannot be detached from the other side of the coin, namely fixed locations, variously termed as *fixity*, *stationarity*, or *sedentarity* (see Kellerman 2006a). "Mobility and fixity, flow and settledness; they presuppose each other" (Massey 2005: 95), even if we assume that "movement [is] becoming a permanent state of affairs" (Bonss and Kesselring 2001). Hägerstrand (1992: 37) noted "that everybody sees the value of mobility but fails to see the equally big importance of stationarity". Thus, we may wonder: "are we experiencing a modification of the balance between mobility and fixity driven by the mobility of people and the circulation of goods, information and ideas?" (Kaufmann 2002: 12), acknowledging that "networks do not have structural effects on territories in terms of the localization of activities", whereas "the usage of networks is sometimes defined by territories" (31). More generally, though, Graham and Marvin (2001: 216) noted: "The linkages between place and technologically mediating networks are so intimate and recombinatory that defining space and place separately from technological networks soon becomes as impossible as defining technological networks separately from space and place".

The traditional distinction between home and work, the two most basic daily fixed locations of individuals, has been blurred in a world of enhanced communications technologies, so that home activities get frequently interrupted by work-related ones and the other way around (Kellerman 2006a). Multiple co-presences have become, thus, possible through the use of the Internet and mobile phones (see Kaufmann 2002: 28; Urry 2000: 71). Such multiple co-presences are mostly noted at home, to a degree of turning the home into a "terminal' (Urry 2000: 72). The interrelationship between daily mobility and fixity is complex: "if we think of space as that which allows movement, then place is pause; each pause in movement makes it possible for location to be transformed into place" (Tuan 1977: 6). But this may also go the other way around, namely that fixity builds up mobility:

"The impetus to motion and mobility, for a space of flows, can only be achieved through the construction of (temporary, provisional) stabilizations" (Massey 2005: 95). Contemporary fixity is relative as, until a few years ago, much of the earth's surface implied full fixity with no terrestrial or virtual reach. Currently there is almost no point on earth which is not connected to electronic communications networks, one way or another. The earth's surface is almost completely covered by satellites which permit satellite telephone communications, and some 90 percent of the populated earth surface permits mobile phone communications.

Daily mobility has to be maintained and controlled through fixed locations, making the mobile always nested within the fixed. Contemporary new technologies have brought about the contrary, as well, namely the fixed nesting within the mobile. New immobile jobs necessary for the maintenance of the mobile may be viewed as such a nesting of the fixed within the mobile. As we will see in the next chapter, mobility is kindled on and fueled by several basic human needs balanced by some basic requirements for fixity. Thus, the dynamic (=mobility) cannot replace the stationary (=location), despite the recently growing significance of the mobile when compared to the stationary. Contemporary individuals find themselves engaged and embedded, more than at any time in the past, constantly and simultaneously, within virtual and physical mobilities and fixities.

Directionality

This term refers to the existence of a defined spatial destination for specific movements or mobilities (see Bonss and Kesselring 2004, Kesselring and Vogl 2004). For pedestrians, Goffman (1971: 28) and de Certeau (1985: 129) noted that some of their mobility is non-directional, such as in shopping trips, or within stadiums, where the very walking and wandering are the important things rather than the reaching of specific destinations. On the other hand, driving implies directional mobility, and this directionality dictates the routing of movement towards reaching of a planned destination.

Directionality is very definite for fixed-line telephone call destinations, since calls are place-specific, with a fixed address for every telephone number. When calling a mobile phone, however, directionality is rather person-specific and not geographic or place specific, as mobile telephone numbers are assigned to persons rather than to geographic addresses. Using the Internet for e-mail correspondence is similar to calling mobile phone subscribers, as messages are sent to persons, who might retrieve them through a computer or mobile phone located anywhere and not necessarily at any base-computer, whether at work or at home. When 'calling' a website, or even more strikingly, when searching for information on a certain topic, the search is completely non-directional, as the searching person is not interested in the geographical location of the hosting computer/server for the transmitted websites, but rather in the information *per se*.

Circularity

Most daily corporeal spatial movements are circular in the sense that they are repetitive movements between the same origins and destinations, e.g. commuting to work or to school; daily or weekly shopping; and regular visits to/by family and friends (see Amin and Thrift 2002: 81-3). Internet surfing may be non-circular in that different sites may be visited at varying times of surfing, but it may also be circular when the same website is used at given times, such as periodical checking of bank accounts. Movements may further be non-circular but still directional, in other words non-repetitive but destination-defined. For instance, travel to business meetings, and shopping at special occasions and places.

Speed

One form of spatio-temporal measurement for both physical and virtual mobilities is checking the duration, or time of movement in space, or over distance, from origin to destination (see Avidan and Kellerman 2004). Hence, enhanced physical or virtual mobilities may imply the speeding-up of movements of both people and information in time-space, respectively. Speed *per se* was viewed as "an irresistable temptation beyond reasonable rational calculation" (Hägerstrand 1992: 35).

Enhanced daily mobility constitutes not only faster electronic transmissions and road traffic, but it may further express social values of speeding-up. "In recent decades, mobility has exploded to the point of characterizing everyday life much more than the traditional image of 'home and family'. Transport …becomes [instead] the primary activity of existence" (Prato and Trivero 1985: 40). The physical experience of car driving speed has been characterized socially as "a pleasure that can even be enhanced by interaction with other cars in traffic, it is certainly a pleasure that can be enhanced by having others observe one's pleasure" (Dant and Martin 2001: 144). On the other hand, calls have been made for short distance travel at lower speeds (Banister 2011).

An increased importance of speed was assessed as lying at the core of *moderne*, notably in cities, as argued by various philosophers of the modern age (see Prendergast 1992: 5; Thrift 1996: 286-9, Sheller and Urry 2000). "Speed is the premier cultural icon of modern societies…Speed symbolizes manliness, progress, and dynamism" (Freund and Martin 1993: 89). Spatially, "it is also the evolution of the city…that comes to organize an existence based upon mobility" (Prato and Trivero 1985: 40). Long before the introduction of the commercial Internet, Virillio (1983: 45) called our era *the age of the accelerator*, and he further claimed that "the military-industrial democracies have managed to transform all social categories into the unknown soldiers of the Order of Speed" (Virillio 1977: 120). However, Hägerstrand (1992: 35) noted that "speed is a relative matter, depending on earlier experience". The constant urge for higher speeds may thus be traced back to early human history with running and animal riding races as exemplary routines. The tremendous importance of speed as a leading social value has to do with the expansive nature of capitalism (see Freund and Martin 1993). Higher speeds for the transmissions of information and resources at large, and of capital in particular, may intensify economic activity, as well as the management of space, and, thus, increase profits (see Harvey 1989).

In daily human practice, increased speed has meant that "mind and eye have become accustomed to seeing and judging partially and inaccurately" (Nietzsche 1983, cited in Thrift 1996: 286). Increased mobility may further cause impressions to be erased by each other, and hence may avoid a deep absorbance of impressions, as well as spontaneous actions. Simmel (1978: 482) commented on the speed characterizing the telegraph and the telephone that makes people "overlook the fact that what really matters is the value of what one has to say, and that, compared with this, the speed or slowness of the means of communication is often a concern that could attain its present status only by usurpation".

The very innovation of the telephone back in 1876 introduced the notions of *immediacy* and *instantaneity* into the world of virtual personal mobility. Whereas the traditional postal service and the technological telegraph implied lagged communications between two parties, the telephone permitted simultaneous two-way conversations fully imitating real-life human interaction. Immediacy and instantaneity have remained special qualities of electronic telecommunications, since they could not be equally applied to corporeal mobility as well, given the very terrestrial nature of transportation. By the same token, the contemporary rush for ever increasing broadband Internet speeds presents a desire to avoid any differentiation in immediacy and instantaneity between vocal connection, on the one hand, and the transmission of the heaviest file of data or streaming pictures, on the other. Similarly, the transmission of video clips over mobile phones has had to await 3G (3rd generation) high speeds of transmission.

Extensibility

Originally defined by Janelle (1973) as "the expansion of opportunities for human interaction" (11), it was redefined later by Adams (1995), as well as by Kwan (2001) to mean "the ability of a person (or group) to overcome the friction of distance through transportation or communication" (Adams 1995: 267). This latter definition refers to potentials for movements but not to their actual materialization. Normally we would assume that people who own a private car, a mobile phone, and/or broadband communications have more extensive extensibilities.

Accessibility

This quality of mobility was defined as "the number of opportunities, also called activity sites, available within a certain distance or travel time" (Hanson 1995: 4). Accessibility, in this sense, is complementary to extensibility. Whereas extensibility refers to the ability to move, or potential movement, accessibility refers to potential locations, or the fixed, to be reached by potential movement. Accessibility, thus, connects between people and locations.

Access is a related but not an equal term. For Kaufmann (2004) access refers to potential movement: "the range of possible mobilities according to place, time, and other contextual constraints" (see also Chapter 4). Access is seldom full and undisturbed. No entry road signs for both pedestrians and automobiles may either prevent access altogether or amount to additional travel or walking, thus increasing travel time and costs. Social barriers, notably within urban contexts, may bring about exclusions of access or reach of certain locations (Jirón 2011). Thus, access may be defined as: "the ability to negotiate space and time to accomplish practices and maintain relations that people take to be necessary for normal social participation" (Cass *et al.* 2005: 543).

Access may further be viewed as reach at several levels (Schutz and Luckmann 1973; Engelbrekt 2011): *actual reach* refers to the space or locations which can be reached by individuals from any given starting point in space, whereas *restorable reach* refers to space\locations beyond immediate reach by individuals stationed at given locations. A third class of reach is *attainable reach*, or locations the reach of which depends on individuals' knowledge and past experiences.

In virtual mobility access barriers are less significant, though one may be denied calling back to a calling party in telephone services if that party's number is hidden. A different source of barriers to virtual access is anti-virus software preventing access to e-mails or websites if they seem infected. Another virtual access barrier may be censorship on websites, stemming from governmental political, religious or cultural apprehansions of free transmissions of ideas, and textual or graphical information (Warf 2011). Sometimes such attempts have turned out less effective then intended given the sophistication of the Internet and its users.

Time-space compression

This trend was defined as "*compression* of our spatial and temporal worlds" (Harvey 1989: 240), or a 'pull' mechanism, induced by contemporary telecommunications. For example, a telephone call taking place between Australia and the UK implies that one of the two parties involved may be awake late at night or working at that time, so that both time and space differences have been compressed. Time-space compression is thus both an outcome and a cause for distanciation, which is the next term we will discuss. By the same token, time-space compression provides for both a separation and a combination between the local and the global.

Time-space compression is not synonymous with Janelle's (1968, 1991) *time-space convergence*, since compression relates to conditions of social space, whereas convergence is a measured index, defined as "the rate at which places move closer together or further away in travel or communication time" (Janelle 1991: 49). Time-space compression reflects power relations. There are those who sense this compression only passively or indirectly, if they are forced to move, or if they are immobile people serving mobile ones, whereas others are in charge of this compression through their handling of the local/global transfers, notably those of capital and information (Massey 1994: 149).

Distanciation

Distanciation was defined as the 'stretching' of social systems in time and space (Giddens 1990), mainly through the development of culture (such as writing) and mobility technologies. Though the term refers to societal trends and processes, it is of significance also to individual daily spatial mobility. Contemporary mobility technologies bring about the 'stretching' of social systems almost to their utmost, thus permitting extended distanciation of reach to users of mobility technologies, whether terrestrial, virtual or aerial ones. One may not be able to access a certain destination, located along a certain route, but the extent of spatial reach, or the distanciation for individuals, is, potentially at least, global. Distanciation may thus be viewed as the geographical separation between local place and global space under intensifying relations between the two. Moreover, an extended distanciation may bring about increased consciousness by Internet users to remote people and places, as part of the *global village* (McLuhan 1964).

Routine daily activities

Golledge and Stimson (1997: 290) defined and contextualized routine activities as follows: "Activities are viewed in terms of routines. A *routine* is a recurring set of episodes in a given unit of time. A recurrence may not apply the same length of time for each activity, because different routines require different periods of operation". The focus of routine activities, as discussed by Golledge and Stimpson (1997), is on activities performed at given destinations. Thus, it was only noted that "also of importance is the mode of transport used to travel between the origins and the destinations that separate the activity episodes" (Golledge and Stimpson 1997: 290). Our focus here is rather on the very daily movements between the locations of activity episodes and, accordingly, on the role of individuals as mobility nodes. In contemporary society daily movements to and between episodes of activity do not necessarily have to be the same or defined as repetitive activities or as 'routines' anymore. This is so notably for virtual mobility, which provides much flexibility for the mobility media used by mobile agents, as well as for changing visited virtual destinations for repeated activities.

The study of routine daily activities in geography has been based on a distinction between two spatial spheres. The first and wider one is *action space*, defined as "an individual's total interaction with and response to, his or her environment" (Golledge and Stimpson 1997: 277). Within this wide geographical sphere of potential action lies the more restricted *activity space*, "defined as the subset of all locations within which an individual has direct contact as a result of his or her day-to-day activities" (Golledge and Stimpson 1997: 279). Such a clear cut differentiation between potential and actual daily spatial spheres of action and activity respectively has become questionable when individuals may perform many activities virtually, under circumstances of globalization, distanciation, and the space of flows, permitting, potentially at least, a global spatial range of both action and activity (see Chapter 11).

Book contents

Following this introductory chapter, the book will be divided into three parts. In the first part, the roots of daily spatial mobilities and their nature will be presented, focusing on needs and triggers for daily mobilities, through the offering of psychological, geographical and political perspectives in this regard (Chapter 2). This will be followed by an exploration of social roots and norms for personal mobility and for personal autonomy in daily mobilities (Chapter 3). The discussion of the various roots of daily spatial mobilities will pave the way for an examination of potential mobilities which eventually may lead to practiced ones (Chapter 4). Finally in this first part of the book the functional nature of daily spatial mobilities will be highlighted through discussions of the question whether terrestrial, virtual and aerial mobilities constitute a joint mobility category or rather three distinct ones (Chapter 5).

The second part of the book will be devoted to modes of daily spatial mobilities, with separate chapters devoted to the examination of terrestrial, virtual and aerial daily mobilities (Chapters 6-8). The third part of the book will deal with three major spatial implications of daily mobilities: urban spatial reorganization in the information age (Chapter 9); mobility terminals, namely central railway and bus stations, as well as airport terminals, notably international ones (Chapter 10); and, finally, global locational opportunities potentially possible by daily virtual mobilities through the Internet, mainly for online shopping, e-learning, home-based business/work, and social networking (Chapter 11).

The book will conclude with a chapter summarizing, first, the book chapters, and followed by separate concluding discussions of the three elements: the daily, the spatial and the mobile agent in daily spatial mobilities. Finally a variety of aspects for both individual and societal management aspects for daily spatial mobilities will be proposed (Chapter 12).

Conclusion

This introductory chapter attempted to present four perspectives for daily spatial mobilities. First, the wider context of mobility at large and other classes of mobility in particular. Second, the rationale for the distinctive interpretations of daily spatial mobilities which are offered in this volume, referring to daily spatial mobilities as routine, two-way displacements, and third were several key terms for the understanding of contemporary daily spatial mobilities in the spheres of their context, movement, and spatial extent. The fourth perspective attempted to show the growing flexibility for individuals as far as movements, and action and activity spaces are concerned.

Within information society each human being may be considered as a node in the global information/communications system. Thus, globalization constitutes a major contextual dimension for daily mobilities in both production and consumption. Virtual global mobility through the Internet is anchored within the space of flows, as well as many jobs, some of which require global physical mobility. Much of contemporary daily communications is performed within networks, both at work and in social-personal communications. Despite these growing mobilities, fixity is still of significance in the creation of mobility and *vice versa*. Movements in daily mobilities may be typified and assessed by their directionality and circularity, as well as by their speed which has become an important social value. The spatial extent of persons equipped with contemporary mobility technologies and able to use them has increased. This is expressed through enhanced extensibility of users and their accessibility to places. Given the nature of mobility technologies they imply growing time-space compression and distanciation for individual mobility agents.

As the discussions so far have shown, the analysis and interpretation of daily spatial mobilities is interdisciplinary, involving mainly concepts and contributions emerging in geography and sociology. Mobility is a process and entity around which much of our lives revolve. The following chapters will attempt to highlight various dimensions of daily spatial mobilities and delve further into them.

PART I
Roots and Nature of Daily Spatial Mobilities

Chapter 2
Needs and Triggers for
Daily Spatial Mobilities

The needs and triggers bringing about daily spatial mobilities are numerous and varied, presenting some of the basics to human existence and functionality. Kaufmann (2011: 35) claims that "we move simply for the sake of moving", and "we move to relax". Viewing mobility as such a primary human need, has led to its acceptance as a basic human right. The *United Nations Declaration of Human Rights* states that "everyone has the right to freedom of movement and residence within the borders of each state". As such, mobility was anchored in North American legislations as well (Imrie 2000).

This chapter will be devoted, first, to two groups of basic human needs and motivations for daily spatial mobilities. The first group consists of three 'push' effects, of an intrinsic psychological nature, leading people to reach out and perform daily spatial mobility: locomotion; proximity (to other people); and curiosity, all three effects being deeply-rooted personal motives for intrinsic mobility. The second group consists of four 'pull' effects, presenting derived demand for mobility, attracting individuals and thus bringing about their daily spatial mobilities: people; places; events; and information/knowledge. All four attracting elements are geographical in nature. They are becoming more and more diversified with the contemporary technology-based extended extensibility and accessibility. The chapter will then explore daily triggers for mobility, such as: commuting, shopping, social ties, information, banking, news, studies, business meetings, etc., all of which reflecting market and social/political forces which may bring about the production of derived mobilities by individuals, rather than their consumption. We will then explore, in the conclusion section of the chapter, some reconciliation between the views of mobility as consumed and produced activity.

Personal 'push' effects for mobility

Man is mobile. He cannot easily stay indoors all day long. He wants to 'exercise his Legs', 'get a breath of fresh air' and feels satisfaction in the mere act of moving, in taking his body and mind from one place to another. We are, after all, descendants from the 'naked ape' who roamed the plains. This quality of travel can be called intrinsic utility (Hupkes 1982: 41).

From yet another, but related, perspective, Hägerstrand (1992: 35) stated: "Physical mobility is a necessary side of our existence as living beings. We have to move in space in order to find resources we need for survival and for social interaction".

Side by side with the basic need for mobility, human beings need some fixity as well for some other basic needs, such as shelter, privacy, and intimacy. Human needs for both mobility and fixity at numerous spheres are specified in Table 2.1, and are discussed in the following sections. We will specifically elaborate on three spheres, each involving a couple of human needs for mobility/fixity respectively: people (proximity/privacy); environment (locomotion/shelter); and information (curiosity/apathy).

Table 2.1 Personal needs, fixity and mobility

Sphere	Fixity	Mobility
People	Privacy/intimacy	Proximity
Nature/environment	Shelter/indoorness	Locomotion/outdoorness
Information	Apathy	Curiosity

Source: Kellerman 2006.

It seems that the basic human needs for fixity and mobility in these three spheres have been interwoven into some balancing in normal human life. As Tuan (1977: 54) put it: "Human lives are a dialectical movement between shelter and venture, attachment and freedom". The fine dialectics and balances between fixity and mobility cannot be stated in a universal way for all human beings. They differ among cultures, with nomads, such as the Bedouins, accentuating spatial corporeal mobility more than residents of Northern cold countries who may experience some winter fixity. Furthermore, the dialectics and balances between mobility and fixity obviously differ from person to person, with some people preferring more fixity than others and *vice versa*. When such interpersonal differences are assessed from the perspective of personal potentials for mobility, they may relate to interpersonal differences in spatial access, personal skills, and the appropriation of those accesses and skills (Kaufmann 2002: 37-9, see Chapter 4).

Historically, until the introduction of mobility technologies, a differentiation existed between *mobility* and *travel*, as far as deeply-rooted needs for individual mobility were concerned. Whereas mobility through walking has always constituted a reflection of basic human needs, travel by animal or cart riding to distant places was historically viewed as must and duty. By the same token, correspondence with partners in remote locations was considered a special privilege. Thus, both travel and correspondence were reserved for particular dignitaries and professions until the introduction of modern transportation media (Bonss 2004). Moreover, movements in traditional societies were not interpreted as mobility, thus creating what Bonss and Kesselring (2001) termed 'movement without mobility', so that mobility did not constitute a value, nor a goal, or an end for itself, nor was it freely chosen, or systematically pursued. As we will see later, contemporary travel, via technologies of transportation and communications, may be viewed simultaneously as reflecting basic human needs to move, side-by-side with its being a rather enforced must, as well as a social value.

The three pairs of human basic needs for mobility and fixity should not be mixed with the constraints proposed by Hägerstrand's (1970, see also Raubal *et al.* 2004) time-geography, namely coupling (activities requiring more than one person for their performance), authority (laws, regulations and norms regulating human activity in time and space), and capability (human abilities to perform given tasks at given places and times). These three constraints refer to human activity *per se*, without a differentiation between fixed or mobile persons and/or activities, whereas here we attempt to identify spheres in which mobility needs are restrained by fixity needs, and *vice versa*.

Proximity

At the social sphere, humans require interaction with fellow human beings. This need for social interaction has been genetically attributed to human attraction to other living organisms (see e.g. Wilson 1984). Furthermore, infants' socialization has been attributed variously to their physiological and neurological maturing (Gesell and Ilg 1943), to social learning and interpretation, as well as to social structure (see e.g. Michener *et al.* 2004). As for the development of human attachment to other individuals, it is questionable, though, whether it reflects or encourages the prosocial (see Giordano 2003). The basic need to interact with other people leads individuals to both virtual and corporeal mobilities.

Interaction with other people may be virtually facilitated through technologies for virtual mobility, notably voice-to-voice interactions through both fixed and mobile telephones, as well as through instantaneous written messages, mainly e-mails and SMSs (Short Message Services). Mitchell (2000: 136) suggested a differentiation between synchronous media, notably the telephone, on the one hand, and asynchronous ones, notably e-mail, on the other. However, the speeding up of communications paces and the introduction of SMS and on-line real-time messaging and chatting have reduced both differences and gaps among communications media as far as their synchronicity is concerned.

Communications technologies permit the development and fostering of social relationships, through voice, text, and visual communications. They may further set the scene for cybersex among communicating partners. Ben-Ze'ev (2004) presented the evolution of romantic and sexual relations all the way from initial virtual contacts using e-mail, real-time exchanges, blogs, and SMSs, through video conversations, to cybersex activities. As such, he considered electronically interactive written communications as a revolution in social-personal relations, notably as compared to traditional letter writing. Virtual written communications permit partners to keep their anonymity and privacy, and thus relations over cyberspace through electronic media may emerge side-by-side with other personal relations in physical space. These options for virtual relations accentuate virtual personal mobilities as an expression of growing individualism in the new modernity (see also Kaufmann 2002).

The wide variety of electronic communications media, including video transmission of callers' pictures and Internet telephone conversations (VoIP), permit not only the development of new types of personal interactions, but also the maintenance and fostering of relations which have been established first in face-to-face meetings, notably if partners cannot physically meet each other. However, despite the growing availability of communications media, and their increasingly sophisticated social uses, social interaction may obviously be more directly achieved through face-to-face meetings. Such meetings require the movement of one or some people to a meeting location, which normally takes place in a fixed location. However, meetings may also be set and take place in moving locations or on the road, for instance on a train or in a shared ride in an automobile.

The basic human need for proximity to other persons is balanced by the human need for privacy, or the need to be at a certain distance or physical exclusion from most or all other people. Privacy may be attained in two ways. First, and in rather fixed space, through 'walled' locations, such as private rooms, offices, or homes in general. Privacy may, though, be sought also in a second form, namely within changing and rather flexible social settings and meetings with other people in space, through the maintenance of a bubble-like invisible *personal space*. By this personal space we refer to the keeping of ever-changing distances from fellow humans depending on the circumstances (personal space in a crowded bus is smaller than in classrooms, for instance) (Sommer 1969, Altman 1975). Furthermore, privacy may be sought also in cyberspace through coded access to one's e-mail, SMS, and other electronic files. At its utmost, the basic need for privacy turns into another basic need, the one for intimacy, either with a selected person such as a spouse, or intimacy of a person wishing to be all alone, in full isolation from other people.

Boden and Molotch (1994) assessed the crucial significance of face-to-face meetings for social contact, in what they termed the *compulsion of proximity*. They stated that proximity via *co-presence*, or via *connected presence*, as Tillema *et al.* (2010) termed it, is of special importance even if varied media for virtual connection are available. Virtual mobility may serve as a substitute for co-presence only if the latter cannot be achieved. Communications via virtual media may eventually require face-to-face meetings in order to deal with the interpretation of electronic ones. Physical co-presence is the preferred medium for human interaction because of the richness of continuous and simultaneous body- and spoken-languages, a combination which is mostly unavailable in virtual media. Partially though signal language exists even in telephone calls, when short or long pauses in speech or response may have implications for the transmission and meaning of messages. Co-presence permits unique ways of conversation, e.g. when one speaker completes the sentences of another, or through laughter, and small talk. Thus, it was found that important conversations are made through face-to-face rather than over the telephone (Tillema *et al.* 2010).

Urry (2002) extended Boden and Molotch's notion of compulsion for proximity, from their main focus on business interaction as implied in their work, to personal social interaction, mainly through air travel (mostly international one, as implied in his writing). Thus, "virtual and imaginative travel will not simply substitute for corporeal travel since intermittent co-presence appears obligatory for sustaining much social life" (p. 258). This same logic would also seem to fit local and domestic social ties, maintained through terrestrial travel in order to achieve face-to-face interaction. However, it turns out that the use of virtual communications media reduces physical, face-to-face contacts (Gershuny 2003, Macdonald and Grieco 2007), applying mainly to the maintenance of strong ties whereas new and weaker ones would need proximity (Larsen *et al.* 2006). Generally, though, all forms of interaction and travel were found to be "of *similar* importance and interconnected with each other" (Larsen *et al.* 2006: 279).

Boden and Molotch (1994) stated rather generally "that there is a hierarchy among forms of human intercourse…Copresent interaction is dominant over other forms of communication in that other forms of communication take their shape through recall or anticipation of copresent talk, rather than the other way around" (p. 277). One may, thus, suggest a potential hierarchy of personal mobilities leading from lagged written exchanges, to real-time vocal ones and eventually to face-to-face ones. As a matter of fact, many written exchanges, as well as vocal ones made by telephone, do not lead to face-to-face proximity, either because the earlier exchanges have not proven fruitful, or because in many cases virtual communication has been fruitful to a degree that physical mobility is not required in order to achieve satisfactory proximity. Human-social interaction may, thus, be viewed in many cases as stratified through the adoption of varied communications technologies. Some specific exchanges, such as romantic ones or business-oriented ones, may begin with written communications, and if fruitful may move on to vocal contacts over the telephone, and only if this phase proves satisfactory then face-to-face contact may be called for.

One may further argue for an opposite hierarchy for the maintenance or protection of privacy/intimacy. Delayed, one-way communications via e-mail or SMS provide for communications while permitting one to keep his/her high level of privacy. In such communications it is up to the receiver of a message whether to respond at all, and there exists some flexibility in response time, as a response may be expected within hours to SMS messages or within days to e-mail messages. In any case, neither physical nor vocal interaction is taking place. A telephone conversation provides for a richer exchange than a written exchange, but at the same time it also implies less privacy, as the interaction is in real time and voices are heard. Physical proximity provides for the richest interpersonal exchange, but privacy is kept to the bare minimum of personal space. When deciding on a form of communication such considerations are taken into account, even if unconsciously and they will not always be admitted when participants are asked to account for the decision-making process for the chosen medium of interaction. Mobility should, thus, be viewed as a careful personal communications process determined not just by constraints and availabilities of time and media, but by conflicting personal needs as well.

Boden and Molotch (1994) further concluded that "copresent interaction abounds under late modernity and, at least among some kinds of actors, is likely more frequent than ever" (p. 277). This effect of increasing face-to-face communications following the introduction of new virtual communications media occurred in the past when the telephone was massively adopted by businesses in the 1920s–30s, bringing about a growth of some 10–20 percent in business travel rather than its decline (Kellerman 1993). Growing travel as stemming from the adoption of new media for virtual mobility attests to users' abilities to do more of the same routine business via virtual communications, and thus leave more time for new business via face-to-face meetings. The introduction and wide adoption of versatile virtual communications media in addition to telephone vocal contacts (e.g. fax transmission and e-mailing) also permits the establishment of new and more sophisticated business on a global scale, because of the ability to transmit documents and interact instantly via fax and e-mail.

Several writers suggested a division of jobs into three distinct categories with differential communications needs (Thorngren 1970, Törnqvist 1970, Olander 1979, Janelle 2004): *orientation* tasks involve the initiation of new projects and decision-making, and they would thus require extensive face-to-face contacts; *planning* tasks represent the implementation of decisions already made, involving less dispersed contacts, and requiring less proximity; and finally, *programmed* tasks include jobs with routine exchanges of information, using mainly telephone and computer exchanges.

Contemporary communications technologies make us receive many written messages, whether through e-mail or through SMS. We may, furthermore, be engaged in a growing number and variety of telephone conversations (through fixed lines, mobile phones, and Internet VoIP). These forms of communication may be rich in content but poorer in human and spatial contextualities. It is questionable how much we are engaged in attempts to imagine these missing contextualities, trying to form a simulated face-to-face interaction. For example, are we frequently trying to grasp the mood of the corresponding person at the specific time of his/her message writing, not to mention his/her facial looking?

Locomotion

Humans require mobility not only for their social relationships but also because of our attitude to Nature and to outdoor environments at large. Human beings need to move in general and in the outdoors in particular. *Locomotion*, as the spatial mobility of individuals is termed by psychologists, develops but is not confounded with chronological age (Liben 1991). Hence, and in a social context, "fundamental is the ability to transcend the present condition, and this transcendence is most manifest as the elementary power to move" (Tuan 1977: 52). Locomotion is also related to curiosity which we will discuss in the following sub-section. "The *locomotion* of the body and its parts offers us the potential to explore and evaluate our environments" (Rodaway 1994: 31). And, "technology also extends the reach of the body and can give us a sense of experiencing a world apart from the body. Here, technologies such as the telephone and television are everyday examples" (p. 32). Our very moving may also satiate a need to relax or "we move simply for the sake of moving" (Kaufmann 2009: 55).

Biophilia, or human relations with Nature, have been dominated by the evolutionary need to survive (Wilson 1984). Thus, the balancing between two requirements which is essential for human survival: the need to move outside, and the constant need for secured indoor shelters for individuals and families alike. Furthermore, movements of any kind require investments of human effort, eventually leading to fatigue which requires shelter. Hence, Hupkes (1982, see also Janelle 2004) differentiated between an *intrinsic utility* of mobility, namely the very benefit from walking, sensing the environment, driving, etc., and a *derived utility*, referring to the value attached to activities at destinations following movements. Both utilities have differential upper values as far as the investments of time, money, and effort in them are concerned.

As several writers have been able to show, travel in natural environments involves information processing. It turns out that such information processing constitutes a most important element for human beings when they choose natural environments for travel. This information processing is expressed in the concern for the complexity of way-finding and in the assessment of the degree of openness of the chosen natural environments (Kaplan and Kaplan 1989, Kaplan *et al*. 1998). Such concerns might apply also to travel at large, not just to travel in natural environments, manifest in the apprehensions that accompany preparations for travel to personally unknown territories. For other people such apprehensions may lead to preferences for staying at home rather than travel, mostly when travel is imposed on somebody by work obligations. The introduction and adoption of GPS devices provides alleviation to some travel anxiety.

It is no wonder that imprisonment has been considered the harshest form of punishment, as it severely limits both the social and the environmental mobilities of inmates. At its most extreme, when confinement in isolation takes place, inmates are confined under the most punishing conditions of privacy, intimacy, and shelter, to a degree of possible psychological and corporeal risk. The need for information, expressed in curiosity (versus apathy) may, thus, become critical, and may carry special significance at times when social and environmental mobilities are restricted, either because of punishment or because of illness and other personal circumstances.

Physical mobility through walking provides for locomotion at its utmost, and to a much lesser degree locomotion may be attained through driving (unless walking is performed at driving destinations). Walking implies exposure to the environment, and if performed outdoors then it further implies minimal or no shelter. Driving, on the other hand, implies staying in a temporary comfortable shelter, at a level of design reminding a living room. Virtual mobility in the form of fixed-line telephone calling provides more shelter, but 'locomotion' in this case might be virtual, or limited by the reception range of cordless telephones. The use of the Internet on a PC (personal computer) is even more restricted as far as physical locomotion is concerned. The use of laptops through wireless communications, and even more so talking over a mobile telephone while walking, may imply instant virtual communications, while being involved in physical mobility, and in the case of walking, communications may be performed under conditions of minimum sheltering. These simple examples of daily communications conducts attest that locomotion and shelter may be mutually interwoven, notably as compared to the sometimes conflicting needs for privacy and proximity.

The need for locomotion is most manifest in the behavior of babies and children in general and in mechanical amusement parks in particular. Such parks, popular all over the world, are mostly based on the pleasure which mobility provides for kids. Furthermore, these parks actually prepare and educate children, at least implicitly, towards all forms and patterns of their future personal mobilities as adults. Carousels, which present horizontal circulative movements, simulate circulative mobility, leaving and returning to the same point, similar to commuting. Giant wheels, or vertical circulative movements, simulate flights. Miniature trains and cars simulate directional mobility, leading to some specific destinations, whereas twisting mountain trains partially simulate non-directional mobility.

Curiosity

The third human need for mobility is for the search and the obtaining of information, including knowledge (Kellerman 2002). "Humans seem to have an inborn tendency to be explorers" (Hägerstrand 1992: 35). Thus, "curiosity aims to explore a space that must still be furnished for us. With questions and gestures more spontaneous than goal-oriented, curiosity explores what it does not yet know and what seems interesting and worth knowing, often for reasons it cannot name" (Nowotny 2008: 3). The human urge for collecting and reaching information is manifest in both virtual and physical mobilities. This statement seems almost obvious as far as virtual mobilities are concerned, as they are all about information transmission and reception. However, physical mobility too provides a wealth of information, which is not necessarily sought for by the moving person, no matter whether the travel objective is for the reaching of people, places, or events. This information refers to the rich spatial contextuality involved in the physical paths used by individuals when involved in corporeal movements. Tacit information exchanges may normally involve some enveloped social contextuality, even in e-mail messages, in the form, for instance, of greetings, because of their informal nature, whereas in codified information reception contextuality is normally excluded. The human urge for information has highly varied motivations, beyond professional, tacit and codified information, or the urge to meet people, places, and events. Basic daily routines involve, for example, a desire to become updated on close family members, resulting in telephone calls, or needs to be updated on other kinds of information, such as flight landing times, leading to a web search on the Internet or to a telephone call.

Of special importance for the understanding of the human urge for the search of information is curiosity, defined as "the desire for information in the absence of any expected extrinsic benefit" (Loewenstein 2002:1), leading to exploratory behavior (see Fowler 1965). In daily life the term curiosity is used in a wider and more permissive sense relating also to urges for information tied to some specific benefit. Curiosity was claimed to have a geographical nature: "When people try to make curiosity explicit, they tend to speak and write geographically. Curiosity is widely portrayed as a desire to encounter the unknown, articulated with reference to various *terrae incognitae*. This is expressed through travel writing and exploration narratives" (Phillips 2010).

Curiosity has become of special importance in contemporary life, as the Internet-based World Wide Web (WWW) has accentuated untargeted curiosities for information and social reaching-out through browsing and networking, respectively. Generally, "more and more means and instruments, mostly but not entirely scientific and technical in nature, are at our disposal to expand the space of our experience" (Nowotny 2008: 3). Traditional libraries in principle may satiate curiosities for information, but the Web provides unprecedented satiation to curiosity with its instant, *in situ*, updated, and extremely varied and multi-sensory information wealth. To paraphrase McLuhan's (1964) famous phrase on media at large, the Internet is not just the medium and the message, but its very availability constitutes simultaneously a motive for curiosity, and a channel for exploratory behavior.

Various origins for human curiosity have been proposed by a number of scholars. Biologically, the natural tendency of the brain to perform as a cognizing organism constitutes a primary source for curiosity. From physiological and psychological perspectives, curiosity may further constitute a secondary drive, stemming from more basic ones such as hunger. In addition, curiosity may be considered a primary drive for resolving certain uncertainties, in ways different than the mind's resolution of other uncertainties. For example in the case of hunger, satisfying it through the eating of one food makes all foods unappealing following satiation, but satiating curiosity regarding one specific issue does not reduce curiosity for other issues. A fourth explanation sees curiosity as arising from a gap between existing and desired knowledge (see Fowler 1965, and Loewenstein 2002 for reviews). One cannot normally reach a full rest from curiosity, but times of relative rest from curiosity, whether one is awake or sleeping, may amount to a form of *apathy* in the sense of incuriosity (see Fowler 1965).

Curiosity is a human drive which relates to information which covers human life in all of its aspects, and hence also the extensive meanings of satiating curiosity for information. Therefore, for certain types of information one might achieve full satisfaction only when physical face-to-face contact is established, even if the travel effort involved in the reaching of such satisfaction is immense. People tend to invest such efforts when close social/family relationships are the case, or when business affairs call for it. On the other hand, getting information on the balance of a bank account is the same whether it is received through the Internet or through walking into a bank branch. By the same token, apathy, in the sense of incuriosity as a required human condition for some rest from curiosity (rather than apathy as a personal characteristic), is a relative condition. The deepest condition of rest/ apathy is achieved during sleep, but sleep may produce unconscious dreams which reflect curiosity. Also, one may rest from one type of curiosity, e.g. resting at home from work-related curiosity, but curiosity may then rise for some personal or any other topics which are not work related.

Personal 'pull' effects for mobility

The basic needs for mobility involve not only the 'push' effects of proximity, locomotion, and curiosity, or an intrinsic demand for mobility, but also 'pull' effects, or a derived demand for mobility motivated by attractions to move, as well (Hupkes 1982). "Travel and communication are derived demands: the goal is to participate in activities at other places" (Vilhelmson and Thulin 2001: 1026). Thus, primary demand for activities in places other than home is conceived of as constituting employment, shopping, entertainment, and any other activities which require travel or communications. These activities produce a secondary or derived demand for mobility by workers or customers in order for them to reach the locations which represent primary attraction for activity.

Proximity at large was differentiated by Urry (2002) as consisting of three motives for proximity which require travel: face-to-face; face-the-place; and face-the-moment (or event). All three attractions reflect a need for proximity to rather specific people, places or events, respectively. In other words, people will travel for three categorical reasons: if they have to meet somebody, if they want to visit specific places, or if they have to participate in specific events. It seems, though, that there are differences in travel flexibility among these three categories, as far as travel timing is concerned. Meeting people sometimes constitutes an urgent must, and sometimes less. 'Meeting places', or touristic visits, might be much more flexible in their timing or necessity as compared to meeting people, whereas participation in specific events is the least flexible, as it requires travel at specific, imposed times. Quite often, though, it is difficult to apply this differentiation given that, as a specific trip may combine two or even all the three categories, as in 'business with pleasure' travel (see Chapter 8). Furthermore, virtual mobility may partially replace business travel via videoconferencing, whereas it can only very partially, if at all, do so for event-based and place-based visits.

All three travel categories are 'enveloped' by virtual travel, as communications means may be used notably before and also after physical travel, since preparations, coordination, and follow-up activities for physical travel are performed via electronic media. The differentiation among the three travel categories for long-distance travel, proposed by Urry (2002), may be applied also for short-distance and daily travels: going to places such as shopping malls in order to touch and try-on merchandise rather than purchase it on-line may be interpreted as face-the-place contacts; travel to work in order to meet colleagues and supervisors rather than using telecommuting may be interpreted as face-to-face communications; and going to events such as concerts, theaters, and cinemas, rather than use TV, video, DVD, and the Internet, may be viewed as face-the-moment travel (see also Chapter 11). In all these categories of daily short-distance travel there has also emerged a contemporary virtual 'enveloping', and in many ways. For instance, getting information about merchandise and cultural events through the Internet; ordering tickets for events over the Internet or through the telephone; and the purchase of products over the Internet after seeing and touching them physically in a store.

Sometimes meeting people, places and events may be superimposed on each other, given the availability of virtual communications. A major and rather frequent example in this regard is the blurring of boundaries between home and work, notably when 'meeting' work-related colleagues over the Internet while being located at a traditionally non-work place, home. At the sphere of social relationships, Licoppe (2004) recognized an emerging pattern of continuous 'connected relationships' through various media of electronic communications, so that "the boundaries between absence and presence eventually get blurred" (p. 136).

A fourth source of attraction in the contemporary Web-based society is information, as a stand-alone source of attraction, in addition to its attraction *vis-à-vis* attracting people, places, and events. A website recommended by a friend maybe an example. The need for daily services such as news and banking is another, and the need to find answers to intriguing questions, even if only words in a dictionary, is a third example. Information and knowledge have always been attractive, but implied in the past the physical reaching out to libraries, agencies and people. The very ability to receive most of one's needed information virtually from almost anywhere has changed the status of information as a factor of attraction for virtual daily mobility, equally important as the traditional attraction of people, places and events.

Personal purchase and production of mobility

As we have seen so far, human beings highly need both mobility and fixity, and, therefore, they might be assumed to be constantly engaged in search for products and services which provide for mobility and fixity, ending up in their purchase or rental. Both fixity and mobility represent, thus, primary or intrinsic demand for a long series of relevant products and services. Buying fixity is first of all expressed in the rental or purchase of accommodation of any kind, normally occurring every several or many years, but it is also continuously expressed in the investment of money, effort, and time in the maintenance of personal housing units, as well as their occasional renovation and refurnishing.

Mobility via technology is too a purchased service involving also the purchase of appliances. One may acquire some potential for personal physical mobility by buying a car, or several cars, or by buying motorcycles or bicycles, followed by constant purchases of a variety of maintenance services. The use of public transportation implies a kind of 'rental' process for temporary transportation services. We further buy virtual mobility services, through the purchase of fixed and mobile telephones, and computers. We obviously pay monthly and variable charges for their use and maintenance to a variety of companies and dealers. Since the suppliers of both physical and virtual mobility services are commercial entities they attempt to have their customers extend the use of their services, promoting customers to make more phone calls and longer ones, or to travel more and buy more petrol. In contemporary societies it becomes, thus, next to impossible to assess specific movements by individuals or even personal mobilities at large as pure expressions of basic personal needs however defined.

Mobilities may be further viewed as a production process rather than as a consumption processes, and in an indirect way. Cresswell (2001a), following Massey (1993), argued for this view by paraphrasing Lefebvre's (1991) famous statement that space is a social product: "(social) mobility is a (social) product" (p. 13). Though Cresswell's (2001a) analysis referred mainly to migrations, Massey's (1993) related to mobilities at large. Differentiated levels of mobility of various sectors or social actors represent "power in relation *to* the flows and the movement" (p. 61), by some very mobile groups, who use their mobility to initiate and control the levels of mobility and movements of others. Thus, "it does seem that mobility and control over mobility both reflect and reinforce power" (p. 62). For example, business people engaged in frequent international travel, or university faculty engaged in intensive virtual travel, may determine the levels of mobility of other people who have to serve their mobility needs through numerous mobility-related jobs as well as other types of employment.

One implication of such power relations is the growing number of people who have to be immobile at any given time in order to serve the seemingly growing virtual and physical mobilities of others (see Cresswell 2001a, Massey 1993). These immobile people may include, for virtual mobility, technicians employed in Internet server farms (Internet hotels), as well as technicians and workers at Internet Service Provider (ISP) firms. For physical mobility such immobile workers may include traffic light control people, as well as traffic policemen. Such immobility requires work in shifts, and if mobilities are assumed to grow, then at any given time more people have to be immobile in order to serve the growing mobilities of others. This trend is part of the growing weekend and night work, where more and more people have to work at 'new' times in order to permit less and less people to enjoy entertainment and shopping at night and over weekends (Kellerman 1991, see also Hubbard *et al*. 2002: 207). We may thus argue that contemporary society facilitates growing *mobility in space* at the expense of *immobility in space* of some people at any time, and, by the same token, contemporary society permits extended *time colonization* (Melbin 1987) through personal consumption at night and during weekends, at the expense of *extended working times* of others.

The power to determine other people's mobilities implies the prevailing of *politics of mobility*, which may operate in a number of ways. An increased purchase and use of private cars, by letting their price be relatively low, decreases the availability of public transportation services, and thus deprives even further weaker social groups who rely heavily on public transportation for their mobility (Massy 1993). The growing buying power for private cars may also be accompanied by car-oriented urban planning, assuming access by private cars to various urban facilities. It further provides for the development of roads, parking facilities, etc. Such planning is not value-free, as it raises the question of who decides which type of mobility infrastructure should be preferred, public or personal (Henderson 2004, see also Hubbard and Lilley 2004). In terms of basic services, Henderson (2004) presents the extreme case of a five minutes movement in order to buy groceries. This may either need an automobile drive along several kilometers, in an automobile oriented urban structure, or alternatively, it may require just a short walk in a more condensed city.

By the same token, the North American *zoning* of urban land-uses, attempting at spatial separation among land-uses, may serve commercial interests, and it may further bring about the need to purchase private cars and their extensive daily use, since shopping and employment centers become geographically separated from residential areas (Urry 2002). The historical cause and effect relationships between the evolution of zoning or automobile-oriented urban land-use systems in the US, on the one hand, and growth trends in car ownership, on the other, require careful historical analysis. Some 50 percent of American households owned a car by 1925 (Kellerman 2006a). Zoning was introduced in New York City in 1916, eight years after the introduction of Ford's T-model, which marked the take-off phase in the diffusion of automobiles into households. By 1926, when over half of American households owned a car, zoning was adopted by 76 cities, and as of that time zoning spread extremely fast, so that by 1936 some 1,322, or 85 percent of American cities, adopted zoning rules (Jackson 1985: 242). Thus, it might well be that zoning and car ownership were mutually reinforcing, so that growing car ownership facilitated the adoption of zoning, which in return called for an even higher adoption of automobiles, eventually leading to daily dependence on them.

Another macro source for mobility as produced and derived demands, rather than as a primary one, was presented by Duranton (1999), who proposed the notion of *tyranny of proximity*, which he advanced through a historical analysis of changing urban economies. At first glance this notion seems similar to the notion of compulsion of proximity (Boden and Molotch 1994), which we discussed with regard to basic, intrinsic, human needs for mobility. However, whereas the *compulsion* of proximity we assessed as a psychological-social drive, the tyranny of proximity is dictated by economic forces. Duranton suggested a *tyranny of distance* (a term originally proposed by Blainey (1996)) for preindustrial cities, followed by a *tyranny of land* in industrial cities, eventually leading to a tyranny of proximity in postindustrial cities. These three forces were presented as the leading economic ones for each era. The development of transportation and communications technologies have decreased transmission and shipping costs and thus reduced urban agglomeration forces, permitting the supply of goods and services ubiquitously, whereas in the past only cities could supply them. On the other hand, and simultaneously with the decline in agglomeration forces, commuting costs have declined very slowly so that urban dispersion forces have remained stable. Under these combined circumstances, cities could be sustained by of the need for proximity. Duranton thus came to a similar conclusion as Boden and Molotch on the relationship between face-to-face and virtual communications: "it even seems that both forms of communication complement each other, so that lower telecommunication costs increase the demand for face-to-face communication" (p. 2183). Telecommunications seems for Duranton to be more suitable for the transmission of codified knowledge whereas face-to-face contacts would be preferred for tacit knowledge (see also Kellerman 2002). Economic changes at the macro-scale may thus sustain and even foster interactions by individuals, constituting mobilities.

Conclusion

In this concluding section we will attempt to reconcile various approaches and views regarding the sources for human individual mobility. We have attempted so far to understand why people move. Mobility has been viewed as stemming from a number of psychological–social motives, consisting of proximity, locomotion, and curiosity, all of which were interpreted as bringing about primary demand for mobility. On the other hand, we noted that mobility may be viewed as derived demand as well, generated by numerous sources, and leading to attraction to specific geographical elements: people, places, events and information. Direct sources for mobility as derived demand are presented by people's activities at locations other than their homes, requiring them to move to various destinations, physically or virtually. Indirectly, a variety of political and economic forces may determine or influence the production of mobilities. First are the very extensive mobilities of certain social sectors and their related power, producing immobilities of others for their own mobilities to function. Second is urban planning which may give preference to specific forms of mobility, and third are economic forces for the sustainability of cities which provide proximity for businesses and business people.

These different sources and effects of mobility seem to coexist and to function simultaneously. Thus, for example, customers of a telephone company may tend to reach out to their friends for proximity reasons, but will be driven to do so more frequently in response to advertisement by their telephone company. As we mentioned already, it is, thus, difficult to identify the exact drives/sources for each human movement, whether it be physical or virtual. Contemporary mobilities tend to be complex, and they become more and more so with the massive adoption of new and more sophisticated technologies. Striking in this regard is the simultaneous uses which can be made of the Internet for work related and personal or social purposes. By the same token, one may go shopping on his/her way home from work, not to mention making a personal telephone call through a mobile telephone, while commuting or while being on a work-related journey.

The decision making process for the adoption of a new mobility medium presents yet another level of choice, albeit on a much less frequent basis as compared to daily decisions on movements (see Chapter 4). One may ask in this regard, what drives a person's decision to buy a PC for home use and have it connected to broadband Internet: does it present a new channel for arousing curiosity and its satiating, or does it rather satisfy a need for a new avenue for social contacts? Is PC and broadband adoption at home merely an act of joining a social trend, or maybe it satisfies the need to keep in touch with work from home? There is certainly a difference in the decision making process for Internet adoption at home between the mid-1990s as compared to the 2010s. During its early years adoption of the Internet at home might have been geared to one or two uses depending on the major user, for instance games for children or work for parents. However, the fast development of the system in terms of the applications it offers has made it relevant to all household members, including children, and has brought about an enormously wide range of applications. In many parts of the developed world the Internet is now a practical must-have. Thus, it might well be that a decision by households to adopt the Internet reflects a need to adopt all its possible uses, or alternatively it may be that adopting some of its possible uses will only build up following connection to the Internet (see Shin and Venkatesh 2004, and see Chapters 7 and 11).

From a societal–political perspective the basic human need to move may be politically and economically manipulated by governments (through planning, regulation, and taxing), by businesses at large (through the location of employment and consumption centers), and notably by the various mobility industries (car companies, airlines, telephone companies, etc.). However, such manipulations are also the case with meeting various other human needs, such as housing (for fixity needs) or food. As we have noted, power manipulations may bring about differential production and consumption patterns of mobilities, as well as growing flexible and permanent immobilities.

The discussions so far of personal needs regarding fixity and mobility presented distinct spheres and categories. However, as our daily experiences may demonstrate, these spheres and categories are interwoven. Thus, fixity and mobility are exchanged on a daily basis between home and commuting as well as among other respective places and rides. People might further be curious to explore a new place in town, and then end up meeting in that new place with some old or new acquaintances, just for the sake of proximity with them. By the same token, we might be annoyed by some weather conditions and thus prefer virtual mobility over walking somewhere. Our life experiences are, thus, simultaneously and constantly embedded in all categories and forms of mobility and fixity, and our complex personal needs structured into them, making them interfolded.

Specific balances among mobility and fixity categories are personal, differing between individuals, depending on personality as well as on individual tendencies and preferences for either more fixity or more mobility. Thus, some persons might be more curious as far as their intellectual curiosity is concerned as compared to others who might be more inhibited in their intellectual curiosity by their very nature. However, the intellectually curious individuals might turn out to be more inhibited as far as their social relations are concerned. Personal requirements for fixity and mobility might further change with age, so that older people would need or prefer more fixity over mobility.

Chapter 3
Personal Mobility and Personal Autonomy

The mobility discourse is deeply connected with the notion of freedom
(Kesselring 2008: 85)

Following our discussion of basic human needs for mobility, expressed through daily mobilities, our focus in this chapter moves to an exploration of social aspects for daily mobilities, through a joint discussion of human needs and tendencies for personal mobility and personal autonomy. If the discussion on the basic human needs for mobility in the previous chapter leaned more towards psychology, the following discussions of mobility and autonomy may expose some of the social roots for daily mobilities.

The objective of this chapter is to assess the notion of personal mobility in light of the wider property of personal autonomy. As we will see, it seems that personal autonomy has rarely been approached from a mobility perspective nor has personal mobility ever been thoroughly examined from a personal autonomy perspective. Mobility has been normally approached from a perspective of freedom, and we will attempt later to differentiate between freedom and autonomy. The joint assessment of personal mobility and personal autonomy is of importance for at least two reasons. First, it is intriguing to examine jointly these two contemporary social trends and values in order to see whether practiced personal mobility truly implies personal autonomy, and/or that it facilitates some additional personal autonomy. By the same token it is of value to assess personal autonomy as permitting and facilitating personal mobility. Second, it is of interest to examine the continuum of mobilities from public to personal (or *vice versa*) in order to find out whether all forms of public transportation actually imply a complete lack of personal autonomy, whereas personal mobilities facilitate or contribute widely to personal autonomy.

The chapter will, first, define and present the notions of personal mobility and personal autonomy, and will then attempt to assess personal mobility in light of the wider property of personal autonomy, questioning the degree of autonomy provided and facilitated by various corporeal and virtual mobility media. The discussion will, thus, begin with a brief presentation of personal mobility (through driving, calling, chatting, etc.) and personal autonomy from various perspectives, complemented by a joint discussion of mobility, freedom and autonomy. These discussions will lead to an examination of personal mobility *vis-à-vis* personal autonomy, identifying levels of personal autonomy by types of mobilities. We believe that personal autonomy and personal mobilities are interrelated, so that personal autonomy implies, among other things, the attainment of personal mobility, and that the availability and implementation of personal mobility may enhance individual personal autonomy. As we will see, personal mobility is not something absolute but it is rather relative, depending on mode of mobility (corporeal or virtual) and the chosen media (public, semi-private or private physical mobility). By the same token the facilitation and affordability of personal autonomy through personal mobility is relational to the mode and chosen media for mobilities.

Personal mobility

Daily individual corporeal mobility can be performed through three major families of media: public transportation (mainly by buses, trains, and boats); traditional bodily-performed mobilities (mostly through walking and also by talking); and through personal mobilities. This latter option of personal mobilities constitutes "technological mobilities" (Graham and Marvin 2001), and defined as "moving of the self by the self" (Kellerman 2006a: 1). Personal mobility, via cars and communications media, has become a complex and sophisticated dimension of contemporary life. Media for personal mobility are mostly technology-rich, undergoing continuous modifications, and permitting fast self-moving of the self. Our interpretation of the mobilities of individuals in light of personal autonomy will focus on personal mobilities, since the very ability of individuals to drive vehicles, or to reach other people and places through their seemingly autonomous self-operation of communications devices, calls for an examination of these actions in light of personal autonomy.

The opposite of personal mobility is not collective mobility, nor a compulsory mobility of the masses from any points of origins to those of destination. The other side of the coin of personal mobility is, rather, public mobility, based on a wide *sharing* of many passengers of the same vehicle, bringing each passenger to her/his destination or close to it. Another possible opposite of personal mobility is personal immobility, to which we will refer later on. Also the opposite of personal autonomy is not collective autonomy, or the autonomy of large scale groups, minorities, or nations, but it is rather heteronomy, or the subordination of individuals' choices and actions to other people or to authorities, notably as far as expression of opinions, religious practices and choice of association are concerned.

As we will see in the following section, personal autonomy has been defined and studied as a social value, trend and need. Personal mobility may be considered a spatial expression, if not *the* spatial expression, of personal autonomy, permitting free and self-governed daily movements in physical and virtual spaces. The fixed-line residential telephone service marked, at the time of its introduction, early beginnings in this regard, by permitting free informational access for housewives and children to other people, having less opportunities at the time for corporeal mobility (see e.g. Fischer 1992). The private car, the Internet and the mobile phone have eventually maximized personal mobilities as expressions of personal autonomies of women and men alike, permitting them to see personal autonomy via personal mobility become an obvious dimension of contemporary life, at least in developed countries.

Personal autonomy

Yet another contemporary societal trend is the growing expectation by individuals to have their personal autonomy expanded, frequently as an expression of growing individualism (see e.g. Friedman 2003). The idea of contemporary personal autonomy refers to persons' autonomy with respect to choices they may make regarding their desires, actions, or character, including freedoms of expression, religion, and association. Both personal autonomy and personal mobility assume the basic value of freedom, to a large degree in its Western democratic sense, permitting people a wide variety of autonomy and mobility, as long as they do not cause risk, damage, or nuisance to others.

The term *autonomy*, stemming from *auto* – self, and *nomos* – law, originally meant "the right to make one's own laws" (Lapidoth 1994: 276). From the perspectives of political science and international law, *personal autonomy* "is normally granted to ethnic, cultural, or linguistic minorities", irrespective of their places of residence within a state, as compared to *territorial autonomy* in which autonomy is granted to a population within specific intra-state regional geographical boundaries (Lapidoth 1994: 280).

The term personal autonomy has been given completely different meanings in both philosophy and the social sciences, pertaining to the personal/ontological status, desires and actions of individuals. In general it was noted that "autonomy is part of the moral basis of personhood" (Young 1986: 25). More particularly, Taylor (2005: 1) differentiated between two historical layers of connotations for personal autonomy: "the Kantian conception of autonomy on which a person is autonomous if her will is entirely devoid of all personal interests," and "a more individualistic conception of this notion, whereby a person is autonomous with respect to her desires, actions, or character to the extent that they originate in some way from her motivational set, broadly construed" (see also Young 1986: 49). Such a latter individualistic personal autonomy requires fellow persons or society at large "to allow persons to form, revise, and pursue their own conceptions of the good" (Taylor 2005: 18-19). These 'own conceptions of the good' normally include freedoms of expression, religion, and association. Furthermore, "the more fortunate a person is as regards the possession of an array of talents and the availability of opportunities, the more paths which hold out promise of fulfillment" (Young 1986: 14), and hence "to be autonomous is to author one's world" (p. 109). In a kind of middle road between Kantian and contemporary individualistic views of personal autonomy, Gill (2001: 105) claimed that *"personal autonomy* implies self-authorship, in that the individual subjects projects and values to critical appraisal and fashions them into a way of life that functions as a coherent whole."

The Kantian approach to personal autonomy, referring to people as devoid of their personal interests, portrays a kind of ideal person and it may fit, even in Kant's own time, only a small number of people. Personal autonomy as freedom from any personal interests may be regarded as a *personal conception* of the self, reflecting high self-esteem, personal values, and/or self-discipline, at times in which individual freedoms, as well as mobility technologies which could both facilitate and express such freedoms, have not yet been developed and realized. Kant's view of personal autonomy could be traced back to the New Testament, as well as to the political thought of the 17th and 18th centuries, but "his views of morality as autonomy is something new in the history of thought" (Schneewind 1998: 483).

On the other hand, personal autonomy in the contemporary and Western sense of individualistic freedoms is rather a *social condition* which is *socially granted and personally applied* by members of society. In other words, a prevailing social norm for wide personal autonomies does not automatically imply that all members of society at all times will tend to take advantage of personal autonomies granted to them. Whereas the Kantian approach is moral guidance for personal reflection and action, the contemporary individualistic approach to personal autonomy may be viewed as constituting a facilitator for action following proper choices by individuals. For Kant it was either that a person had the capacity to have moral laws dominate her/his actions or not, whereas more contemporary interpretations of personal autonomy imply that it is a matter of degree (Friedman 2003: 7; 37-8).

Before delving into degrees of personal autonomy it is important to note that personal autonomy is not just about choices but it is also a matter of action and behavior (Friedman 2003: 4). Autonomous action and behavior normally follow choice and not the other way around, and autonomous action might be aided, expressed and enhanced, by personal mobility technologies. This does not mean, however, that every act of self-mobility expresses personal autonomy. There exists, obviously, a whole range of networks, connectivities and actors who may require mobility for their mere functioning: for example, business distribution networks, and social and political organizations, all of which cannot be considered as directly representing personal autonomies.

The degree of personal autonomy which a contemporary person may attain, or the relationship between one's adherence to the two contradicting norms of autonomy and heteronomy, may depend on a variety of realms, operating simultaneously. Law and social order within a certain national society may be one such sphere, for example as far as freedom of speech is concerned (see e.g. Young 1997: 125-6). A second social sphere might be religion or faith, to which one may adhere, or be obliged to adhere, and which may potentially at least limit, for example, one's social contacts. Another, and third, dimension which may, frequently or occasionally, determine the level of personal autonomy is the family circle of a person, for instance the social and interaction boundaries which may sometimes be set by a spouse, or by children restricting the mobilities of their parents, mainly their mothers'. A fourth one can be language or culture which may limit personal mobilities for immigrants or tourists. In a more general way the degree by which one may realize her/his personal autonomy depends on conditions and social context which might be causal for such a realization (Friedman 2003: 12; 15). A fifth facet of degree for personal autonomy attainment could be a person's own personality, with some people wanting and needing more personal autonomy than others. A sixth element is the personal need to balance between freedom and responsibility (Giddens 1992: 213), so that reflection on desires and wants precedes individuals' choices of their degrees of autonomy (Friedman 2003: 4; 6). A seventh aspect stems from the fact that "freedom is not a given characteristic of the human individual," so that human beings need to expand their mediated experience in order to acquire "an ontological understanding of external reality and personal identity" (Giddens 1992: 47).

The degree to which one may realize her/his personal autonomy in action may also depend on the adoption and use of mobility technologies at large and of personal mobility technologies in particular. In other words, personal autonomy is not just a matter of wants and social contexts but it also depends on the set of mobility technologies available to an individual. These technologies may facilitate more opportunities for the expression of one's personal autonomy, by freely accessing and contacting fellow humans. However, as we will see later, the very use of mobility technologies may be constrained by itself as well. This relationship holds also the other way around: one's very ability to move and to make free use of mobility technologies as much as it is possible depends on the level of personal autonomy granted. For example, the prohibition, until recently, on women possessing driving licenses in Saudi Arabia, as well as Internet censorship in various countries, mainly in Asia and the Middle East, both amount to restricted personal autonomy as perceived by Western standards, and they obviously imply less corporeal and virtual daily mobilities, respectively.

Personal autonomy is not merely a facilitator for personal mobility, but it may also constitute, among other things, the very sociospatial freedom necessary for the establishment and maintenance of self-propelled mobilities as a form of corporeal and virtual expressions. Such extended expressions may bring about expanded and potentially unlimited availabilities of opportunities (see Chapter 11). In order for persons to be considered autonomous, they do not necessarily have to realize an utmost full autonomy or the widest possible personal freedom: "The more one is able to direct one's own life, the greater the degree of one's autonomy" (Young 1986: 8). Partial personal autonomy is the normal general case. As we will see later, a person's degree of autonomy has to be balanced by those of others, so that full personal autonomy cannot actually be achieved (see Young 1986: 110). This same principle applies also to personal mobilities. One's movements have to be balanced by the needs of others for privacy and the needs of others for mobility. This principle has been violated by the use of mobile phones in the public sphere, permitting more freedom to users at the expense of others who are forced to listen to other people's phone conversations (Fortunati 2002).

The degree of one's personal autonomy may also be assessed from a critical perspective, focusing on the status of persons in societal power struggles, with ramifications for mobility as well. Thus, Boltanski and Chiapello (2007) boldly differentiated between strong people (whom they called *great men*) and weak ones (referred to by them as *little people*). If we assume that contemporary society functions through networks, shared by all people, then 'great men' are mobile, exploiting immobile 'little people' (excluding migrations): "Mobility – the ability to move around autonomously not only in geographical space, but also between people, or in mental space, between ideas – is an essential quality of great men, such that the little people are characterized primarily by their fixity (their inflexibility)"... "Geographical or spatial mobility may thus always be regarded as a paradigmatic expression of mobility" (p. 361). Immobile people may lose opportunities for their own development and prosperity, and simultaneously their immobility permits the mobilities of others and their consequent development. In a world of growing universal availabilities of virtual personal mobilities, the full or relative immobility of people may not necessarily reflect just technological constraints, nor would such immobility provide evidence for the existence of social restraints on the use of mobility technologies. Immobility may also be rooted in personal traits and in personal potential tendencies for less virtual and corporeal daily mobilities, side by side with political–economic power relations which may position people's actions.

Mobility, freedom, autonomy

Autonomy in corporeal mobility is a natural and obvious feature of humans, carried out through walking, which has its own social and cultural significances (see De Certeau 1985, Solnit 2000, Ingold and Vergunst 2008). Ancient animal and low-tech mobility technologies implied extended autonomous mobility and distanciation through animal riding and carriage driving. However, the time–space limits, the restricted availability and the efforts involved in such riding and driving have not made these mobility modes reach higher levels of autonomy ubiquitously. It was for technological mobilities introduced as of the 19th century for the expansion of both physical and virtual mobilities that turned autonomous mobility into a powerful social means for reaching long distances fast, permitting more recently global instantaneous reach through virtual communications by individuals. This transition has led some to declare, in light of the wide distribution of the automobile (standing for *auto*nomous-*mobil*ity), that "true mobility can only be achieved autonomously – the distinction between moving and being moved, a passive and decidedly dependent (as opposed to autonomous) state" (Böhm *et al.* 2006: 4).

The extensive innovation of mobility technologies and their wide adoption call for a clear distinction between freedom and autonomy and their implications for mobility. Freedom has a double meaning: freedom from something, e.g.

constraints, slavery, obligations etc., or 'negative freedom', and freedom to do something, such as moving, acting, speaking, etc., or 'positive freedom' (Berlin 1969). Autonomy, on the other hand, is defined as "liberty to follow one's will, personal freedom" (*Oxford English Dictionary Online* 2010). Autonomy constitutes, therefore, something more personal as compared to societal freedom, so that individuals in a society which adheres to principles of freedom may opt for various degrees of their personal freedom, or their autonomy. Freedom constitutes a wider concept which includes personal freedom or autonomy side-by-side with opportunities and choice; rights; and free self-control (Sager 2006).

Geographers tended normally to associate mobility with freedom, rather than with autonomy. Thus, "mobility equals freedom and [that] freedom requires mobility" (Sheller 2008: 25, see also Sager 2006, Blomley 1994, Cresswell 2006b). Spatially, "there is a kind of freedom of mobility inherent in the space itself, or at least in the way it is spatially practiced" (Sheller 2008: 32). But this may be true only for those who enjoy the privilege of access to public space (Adey 2010: 87). Thus, one may argue that the very partition between private and public spaces may have turned into a form of discrimination based on capitalist and patriarchal relations (Cresswell 2006b). Mobility has been recognized as a basic right of freedom in modern society in numerous national and international charters, such as the UN Universal Declaration of Human Rights and the EU Maastricht and Amsterdam Treaties (Cresswell 2006a, 2006b, Urry 2007). The law, then, may produce mobilities (Cresswell 2006b). However, mobility as freedom is not just a matter of legal and social constitution but of individual and cultural aspects as well. Some people might feel less free to move locally if this involves changes in their routines, and some others might avoid international mobility because of linguistic obstacles or cultural prejudices.

Autonomy and mobility have been rarely addressed jointly. On the one hand autonomous mobility has been defined like this: "When we think of personal freedom in relation to mobility we can imagine it in the ideal form as not having one's mobility constrained in any way, and being able to move about and go wherever one pleases at any time" (Sheller 2008: 27). And likewise, "autonomy regarding transport may be translated into the individual right to travel where, when, and how one wishes" (Sager 2006: 470). These are two wide definitions of personal freedom regarding mobility. However, Sager (2006) himself followed his wide definition by stating that "autonomy has to do with the regulation of the private sphere, while privacy is about the ability to keep others outside it" (p. 479). As we stated already, we opt to view autonomy in mobility as the ability of individuals to move themselves as independently as possible. A high level of autonomy in mobility may be achieved through daily personal mobility, and this is practically performed in partial or full privacy, whether through driving cars or making telephone calls.

The very freedom of people to move and the individual level of autonomy derived from it is a complex issue. At the structural level, geometries of power imply that capital and gender relations may determine the degree of access to mobility, making some people immobile for the sake of others to be more mobile (Massey 1993, Graham and Marvin 2001). By the same token our autonomy to buy new cars, computers or telephones, for a seeming satisfaction of our mobility needs, may be distorted by marketing forces (Gartman 2004). At the more individual level, Kaufmann (2002) proposed the notion of motility, which we will highlight in detail in the next chapter. It refers to one's potential mobility, and implicitly to one's autonomous mobility, determined by a variety of personal abilities, status, education, etc. This potential for mobility is materialized by affordances: the car has become the utmost medium for corporeal autonomous mobility (Featherstone 2004), and the mobile phone similarly for virtual autonomous mobility (Gant and Kiesler 2002), but this autonomy is restricted by various factors, such as road congestion (Redshaw 2008), and others which we will discuss later. As a final note, it was claimed that autonomy in mobility does not necessarily imply higher levels of freedom in other spheres of life, as was found in an empirical study carried out by Kaufmann (2002): "Nothing shows that the most spatially mobile people have more freedom in the way they conduct their lives" (p. 58).

Motivations for personal autonomy via personal mobility

As we discussed in the previous chapter, corporeal and virtual mobilities, whether performed bodily or technologically, facilitate the satisfaction of some most basic human needs: *locomotion*, notably through exposure to outdoor environments; *proximity* to fellow human beings through interaction with them; and *curiosity*, satisfied by the acquisition of information (see also Kellerman 2006a). These 'push effects' are joined by 'pull effects' or the attractions provided by any destinations, and which may potentially consist of people, events, places, and information/ knowledge (see Urry 2002). These basic human needs for mobility may be attained by the bare human body, through walking/running, as well as through talking/shouting and listening. The body also enables additional capacities, such as moving the limbs without moving the whole body, so that *movement* at large was defined as "any spatial displacement of the body or bodily parts initiated by the person himself or herself" (Seamon 1979: 3), similarly referring also to the displacement of ideas and objects (Cresswell 2001a: 14), whereas "*mobility* refers to the ability to move between activity sites" (Hanson 1995: 4, see also Kellerman 2006a: 6-9).

Despite the human basic and 'built-in' needs and abilities for mobility it would be too simplistic, and maybe even erroneous, to assume that these needs led at the time to the first inventions of motorized mobility technologies, namely locomotives, followed by automobiles, for corporeal mobility, side-by-side with electronic technologies, the telegraph followed by the telephone, for virtual mobility. Whether one views these primary motorized mobility technologies and electronic appliances as solely invented by individuals or as nesting within social contexts, these inventions were received, first, with much skepticism, irony, and even objection (see e.g. Mackenzie and Wajcman 1999, Hughes 1987). The socially-embedded crucial connections between human mobilities and mobility technologies emerged much later, when the added values of the new inventions were eventually properly assessed and wide clienteles for their adoption and use established. These latter processes have been coupled with chains of demand and supply for the production and marketing of these technologies, fueled by advertising, first mostly in the US (see Kellerman 2006a).

The American emphasis on individual freedoms and frontier development, as basic values of American society since its inception, has contributed to the mass adoption of technologies for personal mobilities in the US, bringing about continuous improvements and innovations in mobility technologies. The massive adoption of these technologies later on in Europe, notably after World War II, was coupled with the emergence of individualism as a major facet of contemporary society (see e.g. Beck and Beck-Gernsheim 2001, Urry 2000). The adoption of technologically-enhanced media for personal mobilities has more recently turned into a global trend, so that the current economic growth in China and India has been coupled with the introduction of cheap family cars and massive adoption of mobile phones.

In contemporary societies the satisfaction of the basic human need for mobility has become inseparable from the daily and continuous adoption and use of technologies. Simultaneously, mobility and technology jointly have been blended with the ever growing social yearning for increased individualism, privacy, and autonomy. This social trend has brought to the front technologies for personal mobilities (see also Kaufmann 2002, Cresswell 2006a). One may argue also the other way around, namely that the very introduction and mass adoption of autonomy-permitting and facilitating technologies, such as SMS and e-mail for written communications, have facilitated enhanced individualism and personal autonomy.

Mobility may further be viewed not as merely something that one does for her/ his benefit or need, but as an act that one may cause other people to perform. In other words, a social condition which may permit some people to enable, cause or compel others to move is a source of power. A simple example is the use of taxis, involving a person who wants to move from a specific origin to a specific destination and who hires a car and a driver for that specific ride, telling the driver exactly of their need or wish. However, personal mobility as personal autonomy refers to one's autonomy to *act on their own, as a mobile person*, and thus the possession of a vehicle is of importance, whereas the availability of money as power for directing a professional driver for a specific ride might be of less importance in this regard.

A more 'hidden' example of power in daily personal mobility is commuting. One may view commuting as either movements by free will or as something forced on a person by an employer and the labor market. In other words, and as we mentioned already, not always does any movement of people by public or personal media imply an action expressing any level of personal autonomy. At the opposite end of such compelled daily mobility may stand movements which are typified by the pleasure associated with the very activity of moving, and even more so by the expectations for pleasure at the destinations of such moving actions. These latter desires typify touristic and recreational actions of mobility (see e.g. Urry 2007).

Levels of personal autonomy by types of mobilities

In her criticism of the definition for personal mobilities, Manderscheid (2007: 165) claimed concerning "the line between self-propelled movements and collective transportations," that moving by car "does not automatically mean that one has to drive the car, the person may only be a passenger, relying on others' capability to move."

This comment raises the question of the status of car passengers from a mobility perspective: should a passenger in a car driven by somebody else be considered as engaged in personal mobility, since she/he rides a private car, or should such passengers be considered as being more similar to passengers in public transportation services, such as trains or buses, since they are passive in terms of their moving of themselves, or, as a third possibility, maybe they should be viewed as passengers in semi-private mobility media, such as taxi services, since they are served by a driver and car exclusively for their own travel? This particular question and its potential answers may be explored through a closer examination of the very nature of personal mobility as an expression of the more general property of personal autonomy.

We may categorize corporeal and virtual mobility technologies by levels of personal autonomy, ranging from technologies and mobility services which require full dependence on fellow mobility agents, through personalized but professionally or socially mediated mobility services (such as taxis), to utmost personal mobility technologies permitting user-autonomy under some basic constraints (such as cars) (Table 3.1).

Riding buses, trains and planes, as well as letter-sending through the postal service, constitute forms of public mobilities, without any personal directing or driving of the mobility vehicles. However, even these public mobility services still imply some personal autonomy in the sense of the timing of each ride or of each letter mailing by the moving persons and these persons' choice of the preferred mobility medium. Letter mailing implies some personalized service as compared to public corporeal mobility services in that letters reach a personalized destination, but this type of geographically focused service does not reflect a higher level of personal autonomy for letter senders since they do not personally move their letters from origin to destination.

Table 3.1 Mobility freedom in public and personal mobilities

Corporeal mobility	Virtual mobility	Level of mobility freedom
Space-constrained automobile driving	Seamless usage of communications media	Personal mobility
Taxis	Courier services	↑
Riding fellow's automobile	Using fellow's communications media	
Buses, trains, planes, and boats	Letters (postal services)	Dependence on mobility agents

Public mobilities may be assessed also from yet another freedom perspective, namely the freedom of passengers within the moving vehicles. Riding public transportation media provides room for much more freedom to read, reflect and work, as compared to driving an automobile. This freedom is even more accentuated in trains, which permit moving about the train cars. It was recently reported for a 2004 survey in the UK that some 34 percent of train passengers read for their leisure; 18 percent were engaged in window gazing/people watching, and some additional 13 percent preferred to work or study, all prevented when driving a car. Two-thirds of the passengers carried mobile phones by then and probably almost all passengers would do just a few years later (Watts and Urry 2008, see also Laurier 2004). Following Adey (2006a), we may thus refer to a relational personal autonomy as far as the mobility experience of individuals is concerned. In any given population we may identify three major sectors regarding the choice between public and private corporeal mobilities: there are those who would always prefer the autonomy to read and rest offered by public transportation, whereas others would always prefer to drive their own cars, and a third group enjoys the very choice between the two mobility modes, making use of the two modes depending on circumstances. However, if personal autonomy means that one would prefer to lead her/his own life independently as much as possible, then driving one's vehicle is personally autonomous, whereas using public transportation or riding somebody else's car is not. Public transportation services express and permit societal freedom to move, but personal freedom in moving oneself is left for driving one's own car.

A significant feature of public corporeal mobility media is that vehicles are shared by many passengers, so that the lack of personal autonomy as expressed through vehicle driving, is coupled with a much reduced level of privacy for passengers. This sharing and reduced privacy does not necessarily nullify totally one's personal autonomy, as far as mobility is concerned, but it implies different norms of behavior in a vehicle shared by passengers who are unfamiliar to each other. The loss of privacy may potentially be compensated by the other side of this coin, namely that public corporeal mobility provides for potential socializing among passengers. Such potential socializing among passengers is obviously much lower when riding a taxi and it may depend on passenger/driver relations when sharing a ride in a private car. Socializing is much more constrained when driving one's own car without any accompanying passengers. Socializing may emerge, though, even when driving a car without passengers, for example when singing along with music on the radio, which may be considered as a kind of hybrid between cars and drivers (see Thrift 2004a). Virtual socializing may be the case for mobile phone communications while driving. In other words, in physical technologically-driven mobility media the more public a medium is the less personal autonomy it permits or facilitates side-by-side with a reverse tendency as far as potential socializing is concerned. These contradicting tendencies stem from the very nature of the vehicles/mobility media, but they do not imply that in general the more personally autonomous a person is the less socializing he/she tends to be!

Things are the other way around for potential socializing when it comes to virtual or virtualized mobility media. The public postal service does not provide any socializing opportunities, other than a written one between specific letter senders and receivers, which is the purpose of such a specific communication, but this communication involves significant delayed time intervals. The telephone, on the other hand, permits multi-participant conference conversations, and the Internet permits numerous socializing opportunities through social networks of all kinds, while keeping one's privacy and personal autonomy as to the choice of those who are being communicated with, and those with whom full self-identities are exchanged. Such interactions may counterbalance traditional corporeal family and social ties, and they may bring about new social face-to-face meetings and relations (see Ben-Ze'ev 2004).

There are two intermediate forms of mobility, positioned between public and personal ones, namely personally-hired professional mobility services, and joining somebody in her/his personal mobility vehicle on a family or friendship basis. With this classification in mind, carpooling is, on the one hand, similar to a hired mobility service (such as a taxi), but, on the other hand, it involves colleagues/ friends rather than randomly hired professional drivers. The most frequently used services among the hired mobility forms are taxis for physical mobility, and courier services for the pick-up and delivery of letters from personal addresses, for virtual mobility. Joining fellows' mobility media occurs when somebody is riding a car driven by a family member or a friend, or when one makes use of somebody else's communications devices (e.g. mobile phones) or e-mail address. All these forms of mobility amount to sharing preferred timing, duration, routing, and delivery for the mobile persons, as compared to respective public mobility media, but they further imply the facilitation and provision of only partial personal mobility and autonomy. A passenger in intermediate mobility media might get from an exact point of origin to an exact point of destination using a car, but her/his status is that of a passenger, so that the mobility vehicle is not her/his own property and the driving, in the case of physical mobility, is performed by somebody else. It seems that socially-based riding is closer to public transportation than to riding a hired vehicle, since the very use of mobility services depends on the goodwill of the driver/owner, which may also frequently imply the provision of a partial service only, as far as pick up location and the reaching of an exact destination are concerned. On the other hand, however, when hiring a mobility service, then its availability is normally taken for granted, as well as full pick up and drop off services.

So far we differentiated mobility types by the level of autonomy which they permit/facilitate. We will turn now to a comparative examination of the two basic mobility categories, corporeal and virtual, in an attempt to assess the levels of autonomy which they facilitate. As the following discussions will show, it seems that virtual mobilities at large provide for higher levels of personal autonomy, though this distinction may not be so clear cut, and corporeal mobilities have some autonomy advantages over virtual ones in some respects.

Driving a car or using one's own communications media imply autonomous mobility in that persons direct themselves corporeally or virtually from origins to destinations. Virtual mobility seems more impressive in this regard as one may move any type of information in any format to any other connected node or person worldwide, reaching their destinations instantly. Being autonomous in the sense of car driving, and even more so in the use of telecommunications media, still implies user reliance on heavy and rather complex 'behind the scenes' infrastructures facilitating the autonomous driving/use of personal mobility media. The time and money savings in electronic virtual mobility, as compared to letter sending, are much higher than the savings in time friction and space and distance frictions in personal mobility through car driving, as compared to those of public transportation.

Both driving and the use of communications media do not provide for or permit the achievement of ultimate full personal autonomies. In driving it is obvious that one cannot move from one point in space to another one in a straight and shortest possible route, free of any obstacles; even expressways have their own spatial contours. Driving is rule-bound, and it involves, among other restrictions, no-entry and one-way streets, and it further involves traffic jams, which may normally not be the case in virtual communications, though limited telephone systems, based on copper cables, as well as narrowband Internet connections, may cause virtual traffic jams as well. Massive traffic generated by many drivers materializing their personal mobility implies heavy environmental side effects, mainly air pollution which is brought about by each car in operation, which is not the case with virtual mobilities, other than the radiation of mobile telephony antennas, and the implications of the high power use by Internet server farms, both of which are side effects of the systems at large not produced by each user individually. The congestion and pollution caused by corporeal personal mobility are the cornerstone for policy attempts to balance between personal autonomy via personal corporeal mobility, on the one hand, and public mobility through public transportation means, on the other.

Virtual mobilities involve social rather than material restrictions on the way to full personal autonomy. Thus, there are occasions in which the use of mobile phones is not permitted, and not everything can always be expressed over Internet websites.

The restricting of physical and virtual mobilities is not a novel idea. In the pre-technology era spatial privacy had to be honored by not transgressing private property and not entering someone's home without explicit permission, which was also true for family rules regarding permission to enter specific rooms within a family's home. By the same token, not everything could be spoken out loud in private conversations or in public gatherings. Thus, some balancing has always been sought between one's personal autonomy as applied to free mobility, on the one hand, and fellows' personal autonomies as expressed in their home space and in their personal honor, on the other. Since all members of society have always wanted to enjoy reciprocally both their mobilities and their possessions, privacy and honor, social norms and rules have developed to balance between mobility and privacy, as well as between the static or fixed (spatially) and the mobile (physical and virtual).

Driving one's car normally implies that a person stays within her/his privately owned property, guiding it his/her way as much as possible under traffic regulations and conditions. However, this property, the car, is not as guarded as one's home normally is. Police may routinely check one's automobile ownership, and under circumstances of extreme violations of traffic laws numerous countries permit police officers to confiscate one's car temporarily on the spot. Such a procedure is impossible regarding one's immobile home as well as one's mobile communications devices while on the road. The personal nature of the latter is becoming more and more accentuated as compared to people's automobiles, in that these devices increasingly contain highly personal information, such as messages, lists of telephone numbers, previous calls made, personal pictures and music, and more. However, these devices are not checked by police on the road and cannot be confiscated. The car is considered as a potentially harmful appliance to the bodies of both drivers and other road users whereas the mobile phone is not. This difference between cars and communications devices also works the other way around. Cars normally do not record origins and destinations of trips made in them, whereas communications media do. Thus, though ownership rights are more honored for communications media as compared to automobiles, and therefore provide higher levels of personal autonomy in this regard, they are much more vulnerable to screening, in violation of one's personal autonomy, usually through illegal attempts to do so.

Personal autonomy through personal mobilities is not only an 'active' matter, namely that mobility action may permit actors to enjoy higher or lower levels of personal autonomy. Personal autonomy may also relate to the 'passive' side of personal mobilities, namely the degree to which people may permit others to access them. Thus, one may permit others free corporeal access to their homes, to their e-mail addresses as well as to their telephone lines, or persons may filter access to them by using messaging systems or by blocking access altogether. People may use such filtering and blocking options permanently, temporarily, or in discriminating ways, permitting access to them through specific communications media. These options have become widely used and have become more sophisticated in nature through recent continuous enhancements of virtual mobilities, thus permitting the emergence of a sophisticated 'passive' personal autonomy in the form of the assurance of privacy. We will return to these differential levels of access in the next chapter focusing on potential mobilities.

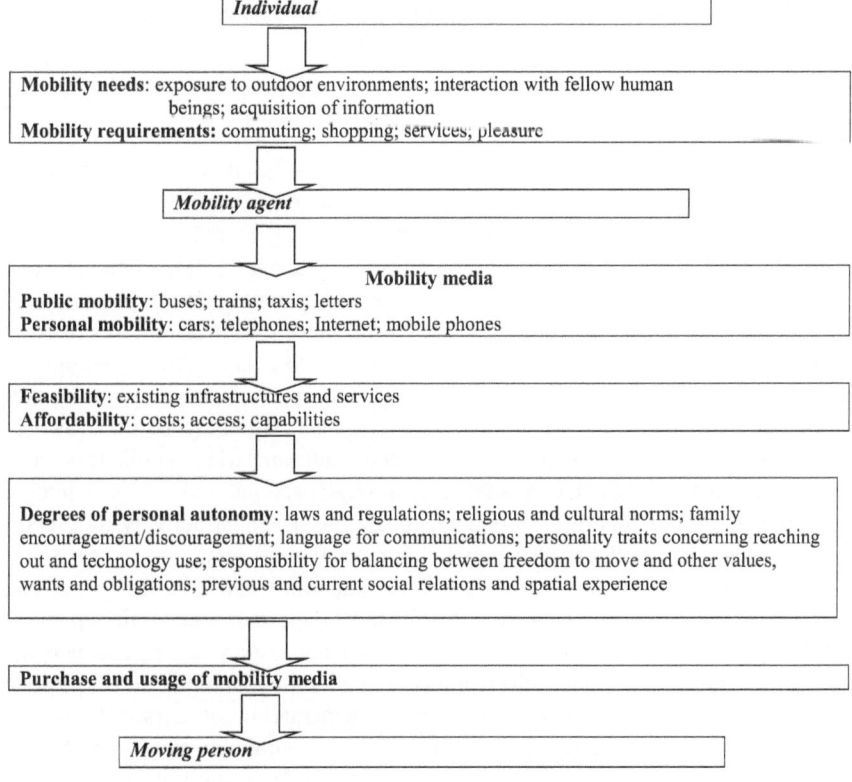

Figure 3.1 Individual mobility motivations, choices and considerations

Our elaborations so far may permit us to portray how individuals become, first, mobility agents, and eventually moving persons (Figure 3.1). The basic human need for mobility coupled with daily requirements for travel and communications turns all human beings into mobility agents. In considering the various media for public and personal, corporeal and virtual, mobilities, people take into consideration their feasibility and affordability, side-by-side with the degree of personal autonomy they would like to experience. These considerations lead to the possible purchase of personal mobility appliances and services, as well as to the much more frequent decisions on their very use for specific movements. These decisions yield moving persons, corporeally and virtually, enjoying differing levels of personal autonomy on any given corporeal and virtual trip. In the following chapter we will attempt to extend these notions one step further, namely leading the personal needs and the social circumstances of individuals' mobilities towards potential mobilities.

Conclusion

Personal mobility as personal autonomy is a specific condition within the wider societal freedom to move. Thus, being free to move does not necessarily imply full or partial autonomous mobility. The attainment of personal autonomy through personal mobility involves a heavy use of mobility technologies, so that personal autonomy, as a social value, is not only expressed spatially through people on the move, but it is further embedded in the spatial mobility infrastructures of roads and cables, and even more so in mobility artifacts like automobiles, laptops and mobile phones. Even when such vehicles/appliances stand still they tell by their very nature of the personal mobility and personal autonomy of their owners/users. Since cars can be easily turned on and communications appliances easily carried along by users, personal autonomy through personal mobility is an integral experience of contemporary mobile social life for all, either actively by users of personal mobility media or passively by temporary non-users.

Personal mobility constitutes personal autonomy and frequently personal autonomy is personal mobility. Probably the most important social trait of personal mobility is its built-in interrelationship with personal autonomy, so that personal autonomy implies, among other things, the potential attainment of personal mobility, and that the availability and implementation of personal mobility may enhance human personal autonomy. Personal autonomy is not an absolute condition, but it is rather a relational one, at least as far as mobility is concerned, depending on the mode of mobility (corporeal or virtual) and the chosen media (public, semi-private or private physical mobility). One may prefer the autonomy to think, read, and even move about a vehicle while on the move, whereas others would prefer to be autonomous as much as is only possible through driving their own vehicle, alongside many other car drivers. By the same token, the facilitation and affordability of personal autonomy through personal mobility is relative to the mode and chosen media for mobility. Thus, SMS communications may provide more personal autonomy in its flexibility of instant and silent personal contact, as well as its universal spatial accessibility, as compared to less flexible fixed-line telephones, and even voice mobile telephony involving the potential to be overheard by others.

Personal mobility in the technological era is a condition in which persons find themselves on a daily basis. Beyond daily requirements to move corporeally and virtually for work and pleasure, some people just like to drive, or enjoy making use of their telephones and the Internet. Do such personal likings imply that those who make more intense use than others of personal mobility media have higher degrees of personal autonomy? Not necessarily. It all depends on the destinations and uses/movements made and on the mindset of users. Thus, expanding one's social and personal circles may be achieved through constant, but not necessarily continuous, use of telecommunications, and by the same token one may use ICTs continuously but for just a few or shallow contacts and/or website surfing.

A form of mobility not discussed in this chapter is flight, which will be discussed in Chapter 8. Going to foreign countries, notably through flights which may take passengers, within a few hours, to a completely different cultural setting, may provide passengers with a sense of personal autonomy; even though the flight *per se* might not involve any real expanded personal autonomy, given that airlines constitute a mode of public rather than personal mobility (see Kellerman 2006a).

The examination of personal mobility in general, and of personal mobility in light of personal autonomy in particular, calls for potential examinations of mobilities *vis-à-vis* other social values, notably those values which represent the other side of the coin of personal autonomy, namely societal concerns of individuals who might be seeking high levels of personal autonomy. Social responsibility constitutes such a relevant social concern and value, and it applies, for instance, to the conduct of drivers and mobile phone users in public places, to the behavior of passengers on public transportation vehicles, to the contents of widely distributed e-mail messages, as well as to the production of personal websites and blogs. A question in this regard which deserves separate treatment is: are there any new norms of reduced social responsibility emerging in favor of expanded personal autonomy in the contemporary world of expanded personal mobilities?

Personal mobility constitutes simultaneously and jointly a social and spatial subject matter. Empirical examinations of the interpretations of personal mobilities through personal autonomy as proposed in this chapter require more attention to people than to spaces and places. An intriguing path for empirical study could be an attempt to understand the various uses made of mobility media in light of the aspects which were proposed for the existence of degrees of personal autonomy: law; religion; family/spouses; language/culture; personality; sense of responsibility; and social relations and spatial experience.

Chapter 4
Potential Mobilities

So far our discussions have focused on various human needs and social meanings of daily spatial mobilities. In this chapter some of the concepts and ideas presented before are contextualized and incorporated into the understanding of potential mobilities at large, but those for daily spatial ones in particular. Furthermore, as we will see, the very notion of people's potential mobilities refers largely their ability to move, which is based on their personal needs and abilities, as well as on societal circumstances for daily mobilities as presented in the previous chapters.

The study of mobilities has flourished during the last two decades, and by the very nature of mobility as a human action, much attention has been given to the various mobility activities and options as well as to their social and spatial significances. Less attention has been devoted to the potential of *homo viator* [mobile Man] (Eyerman and Löfgren 1995) to move. It is of importance, though, to understand potential mobility, notably before moving to the discussion of the operations of daily spatial mobilities through the numerous mobility media in the next part of the book, and for several reasons. From the perspective of mobile people, or mobility actors, knowing of their potential mobilities may assist in the interpretation of their specific movements or practiced mobility at large. Furthermore, and from a conceptual perspective, potential mobility is one side of the mobility coin, coupled with its other side of practiced mobility, so that the very study of mobility at large has to include an examination of potential mobilities as the preparatory phase towards practiced mobilities.

The term potential mobilities has, therefore, a double sense: the potential *for* mobility of individuals, and the potential *of* mobility of eventually practiced movements/activities prior to their occurrence. The potential of mobility is mostly dependent on the potential for mobility of potential mobility actors. If a person is capable of being mobile then the chances of a certain movement to take place are higher.

The objective of this chapter is to highlight the issue of potential mobilities, by first presenting some possible basic terminology for potential mobilities, followed by a critical review of motility as potential mobility, and continuing with an attempt to put forward elements for potential mobilities at times of wide availabilities of mobility technologies (Kellerman 2012). These elements for potential mobilities include: definitions and meanings for potential mobilities; a discussion of active and passive potential mobilities; and an examination of potential mobilities in light of practiced ones. These discussions will permit to suggest a simple model for potential mobilities focusing on the accumulation of mobility needs by individuals, their access to mobility media and their competences to make use of them. These elements may lead to an appropriation process which may bring about various modes of mobilities (such as immobility, accessibility or passive mobility, as well as several forms of active mobility).

Possible basic terminology

Several basic terms have been, or can be, considered for human potential mobilities. The most well known term, *motility*, was taken from biology, referring originally to the very ability of animals to move, but used mainly for single-celled and simple multi-cellular organisms, as compared to animals' locomotion. Motility, in its biological sense, refers also to the ability of organisms to move food within their bodies (see e.g. Biology-Online 2010). The opposite term, *sessility*, characterizes mainly plants, referring to their inability to move (see e.g. Biology-Online 2010).

As we will see in the next section, the extension of the notion of motility to the potential mobility of humans was proposed by Kaufmann (2002), though it was used already before, albeit less explicitly and with some differing connotations, by Bauman (2000), Mol and Law (1999), and Virilio (1992, 1998). The extension of the biological term of motility to the potential mobility of contemporary technology-rich humans blurs the distinction between locomotion and motility. It further implies, at least implicitly, that human potential for moving through natural means, i.e. walking and running, is similar in its basic nature to the human ability to move through most powerful man-made technologies. Moreover, the original biological term of motility does not refer to virtual mobility of organisms via sound making and other signaling by animals, whereas human technology-based mobilities consist of both corporeal, and virtual sound and signal-based, mobilities (Kellerman 2006a).

A second possible term for potential mobilities could be *movability*. However, this term has both active and passive connotations. Actively it refers to "the quality or condition of being movable; mobility" (Oxford English Dictionary Online 2010), whereas passively it means "capable of being moved" (Merriam-Webster Online 2010). Potential mobilities may refer, as we will see later, to potential active and willing human mobility as well as to passive potential mobility generated by others, but not to humans being literally moved by others, a movability which characterizes goods and information.

Potential mobilities could have, alternatively, been considered as a major dimension of a wider notion of human *action potential*, or *action capacity*. These terms would also have included, in addition to mobility, human potential actions via all five senses, such as viewing, touching, hearing, and smelling. Like mobility, some of these capabilities have been extended tremendously through technologies, such as listening, viewing, and speaking abilities immensely developed through ICTs. It turns out that the social sciences have not focused on human potential action in general under the terms action potential and action capacity. In biology, though, the term action potential is used for certain specific short lasting events in which some rise and fall of a cell's electrical membrane potential may occur (Barnett and Larkman 2007).

Given the problems associated with these three terms we preferred to use the term *potential mobilities*, avoiding any direct or metaphorical use of biological and other terms. However, since the term motility has been proposed and developed for potential and other mobilities we will critically review it first before our attempted explicit development of the notion of potential mobilities. This critical review of motility will highlight several aspects of potential mobilities.

Kaufmann's motility

As mentioned already, it was mainly for Kaufmann (2002) as well as for Kaufmann and his partners (Kaufmann *et al.* 2004, Flamm and Kaufmann 2006, Kaufmann and Montulet 2008) to develop the notion of motility for human potential mobilities and to expand on it. Though this notion was discussed in various mobility texts published in recent years (e.g. Kellerman 2006a, Adey 2010), and it was further incorporated into Urry's (2007) mobilities paradigm, it was applied partially in only one empirical study (Kesselring 2006), other than Kaufmann's (2002) own.

Moving through the various definitions for motility proposed by Kaufmann and his partners, sometimes more than one definition within a given essay, it is not completely clear what are really the definitional boundaries for motility. Altogether some seven definitions for motility were found:

1. "The way in which an actor appropriates the field of possible action in the area of mobility, and uses it to develop personal projects" (Kaufmann 2002: 3).
2. "Motility can be defined as the operation of transforming speed potentials into mobility potentials" (Kaufmann 2002: 99).
3. "Motility can be defined as the capacity of entities (e.g. goods, information or persons) to be mobile in social and geographic space, or as the way in which entities access and appropriate the capacity for socio-spatial mobility according to their circumstances" (Kaufmann *et al.* 2004: 750).
4. "Motility, the actual and potential spatio-social mobility" (Kaufmann *et al.* 2004: 753; see also p. 754).
5. "Motility can be defined as how an individual or group takes possession of the realm of possibilities for mobility and builds on it to develop personal projects" (Flamm and Kaufmann 2006: 168, see also Kaufmann, 2009: 58).
6. "Motility may be defined as the manner in which an individual or a group appropriates the field of possibilities relative to movement and uses them" (Kaufmann and Montulet 2008: 45).
7. "Motility is both mobility and one of its components" (Kaufmann 2011: 40).

Definitions 2 and 3 relate to motility as potential mobility, whereas the other five refer to motility as a very wide spectrum of mobility, including both potential and practiced mobilities. Moreover, whereas five out of the seven definitions refer to motility, whichever way defined, of humans only, definition 3 attributes motility also to goods and information, entities which by their very nature are being moved, as compared to humans' ability to move autonomously, and in addition, possessing the capability of having goods and information moved. This latter definition, thus refers to motility as movability and it treats goods and information as entities with an ability to access and appropriate and implicitly also as being able to make a choice for mobility (Kaufmann 2002: 53). Partially, at least, one may assume that these differences in the definition of motility may reflect an evolution of thought with regard to the very nature of motility.

Kaufmann and his partners developed several aspects for motility, two of which we will critically review here: motility factors and motility as capital. Based on Lévy (2000) and Remy (2000), Kaufmann (2002) and Kaufmann *et al.* (2004) proposed three factors which determine a person's motility: access; competence or skills; and cognitive appropriation. All these three factors are assumed to be interdependent with each other. We will discuss now each of these three factors.

Access refers to the availability of mobility possibilities for individuals, including *options* for technology adoption and their pricing, as well as other *conditions*. Thus, access was originally proposed as relating to less frequent decision making by individuals regarding the purchase and adoption of mobility technologies, such as the purchase of automobiles, the Internet and mobile phones. This is important to realize since the third motility factor, that of appropriation, relates to routine and continuous daily decision making by mobile agents on possible mobility practice. Access here constitutes a social factor which is not synonymous with the spatial term *accessibility*, defined as "the number of opportunities, also called activity sites, available within a certain distance or travel time" (Hanson 1995: 4, see Flamm and Kaufmann 2004) (see Chapter 1). Kaufmann's definition is in line with a growing use of the term 'access' beyond its spatial connotations, referring to public access to resources and opportunities (see Rifkin 2000). Like spatial accessibility, people's access to mobility technologies is seldom full and undisturbed.

The scope of access to mobility options and technologies may sometimes be a matter of geography and culture. Thus, in the Netherlands, for example, access to mobility options will almost always include bicycles, and in Amsterdam frequently also private boats. In Rome, mobility options may frequently include light motorcycles (*motorini*), and in Venice boat-buses are the only mode of transportation about the city. As for virtual mobility media, Koreans and Japanese tended to be the first global customers to adopt *smartphones* which combine various modes of communications (visual and audio; Internet and cellular, etc.).

Access to mobility media cannot be considered as merely a matter of purchase and adoption, or major changes in the personal arsenal of mobility media of mobility actors. Access may further, not instead, be a daily factor involving, for example, interdependence conditions, notably among family members. This may be the case, for instance, if a couple has only one car at their disposal, bringing about their taking of turns in its daily use, forcing one partner to use public transportation each day. Similarly, a couple or parents and children may share the use of home computers, so that access or potential virtual mobility of each family member may change throughout the day.

Competence refers to mobility *skills* and abilities, at both the individual and societal levels. This factor accentuates the nature of motility as varying from person to person, on the one hand, but also having a more general societal nature, on the other (see Kaufmann 2002: 40). It includes three groups of skills: physical; acquired; and organizational. Thus, *physical ability* is one's ability to move entities from one place to another, as well as one's own self- or personal mobility. It excludes the ability to move virtual information, since the moving of information does not require any physical carrying of information other than possibly the carrying of an information terminal (e.g. mobile phones and laptops). *Acquired skills* relate to mobility licensing, notably driving licenses, and to language knowledge as mobility means, e.g. for virtual mobility over the multilingual Internet. All these skills relate to one's personal abilities, but less to societal rules and regulations regarding mobilities. *Organizational skills* include the synchronization of activities and planning, both of which might be personal and involving several people. At the personal level all the three competences (physical, acquired and organizational) may change along one's life cycle.

Motility seems to depend also on a third element, in addition to access and skills, namely social and cultural contexts and structures within which people live, stretching beyond laws and regulations pertaining to the use of mobility media. These may include the factors which we outlined in the previous chapter regarding personal autonomy: religious and cultural norms pertaining to the use of mobility media, notably virtual ones; encouragement or discouragement of mobility by family members, especially spouses; personality traits concerning reaching out to other people and places; one's sense of responsibility regarding balancing between the freedom to move and other values, wants and obligations; and previous and current social relations and spatial experience as promoting or blocking one's mobility. As we have noticed already, low motility of persons may attest to enforced immobility through power relations which may bring about the mobility of stronger elements in society at the expense of weaker ones (see Boltanski and Chiapello 2007, Massey 1993).

Appropriation refers to the ways in which mobility agents evaluate mobility options in light of their personal levels of access and competences, weighed by aspirations, motives and needs, before each potential movement. It includes an element of 'choice of motility' and compromise (Kaufmann 2002: 53, see also Kaufmann and Montulet 2008), not just existing or imposed access and competences. Of the three motility factors the appropriation process might be the most complex one, as it involves, at least indirectly, various elements in addition to access and competence. First is human motivation to move as discussed in

Chapter 2, including the intrinsic 'push effects' of locomotion, proximity, and curiosity, which may differ from person to person (Kellerman 2006a), and the 'pull effects' of events, places, people and information (see Urry 2002), side-by-side with derived needs for daily movements, such as commuting, shopping, etc., which may be considered as triggers for specific movements. All these three motivation groups may contribute to individuals' access levels and their acquired competences/skills for mobility. At the appropriation phase, mobility needs are weighed in light of the multitude of access and competence elements, as well as in light of the social and cultural contexts of mobility actors, possibly yielding several options of mobility modes for a specific planned movement, eventually leading to the choice of a specific mobility mode, or alternatively to avoidance of mobility at all.

In an empirical study of mobility Kaufmann (2002) found that: "Nothing shows that the most spatially mobile people have more freedom in the way they conduct their lives" (p. 58). It is, thus, of no less importance to realize the spatial consequences of one's motility, or one's capacity and propensity to move, involving major decisions on technology adoption, side-by-side with cumulative daily decisions on travel distance and travel means. Increased mobilities via transportation, and even more so through communications and information technologies, result in *distanciation*, or the 'stretching' of social systems in time and space (Giddens 1990). The spatial extent of distanciation may depend on people's motility which may differ from person to person. The same interpersonal motility differences may potentially apply also to the use of the Internet which permits an *extensibility* of human beings, in that the instantaneous and global reach offered by the Internet relaxes time-space constraints concerning mobility and activity space (Adams 1995, Kwan 2001). The very use of any wide-reaching and speedy medium such as cars and the Internet does not automatically imply distanciation and extensibility of one's activity space, since one might opt to reach out only within certain, rather limited, local, regional or domestic spatial spectra.

Bourdieu (1984, 1987) proposed a distinction between four categories of *capital*: *economic; social; cultural*; and *symbolic*, all of which constitute resources which can be acquired, transmitted and converted. Urry (2007) proposed a fifth form of capital, *network capital*, relating to practiced mobilities at the societal level: "Such network capital is not to be viewed as an attribute of individual subjects. Such capital is a product of the relationality of individuals with others, and with the affordances of the 'environment'" (p. 198). Motility was claimed to constitute an additional form of capital (Kaufmann 2002, Kaufmann and Montulet 2008), but its relationships with other forms of capital received contradictory assessments. On the one hand, "motility represents a form of capital that is independent of other forms of capital traditionally taken into account when analyzing social position and other indicators of social integration" (Flamm and Kaufmann 2006: 184), but on the other, "motility forms theoretical and empirical links with, and can be exchanged for, other types of capital" (Kaufmann *et al.* 2004: 752). It seems, though, that the latter statement is more realistic. Consider the following two simple examples. A person opts not to travel locally in order to receive a package, and instead hires a delivery service. That person's ability to move or motility has been turned voluntarily into immobility and is exchanged for money used to hire the service of professionally mobile workers. Or, as an additional example, a Professor invited to lecture in another city sends instead an advanced student, trusting the student's knowledge. The Professor's motility is exchanged with the knowledge and mobility of the student. This latter example is more complex than the first one regarding motility as capital, since it includes not only motility in terms of an ability to move and the time and cost of travel involved, but also a substitution of knowledge, that of the Professor with that of the student. Thus, a high level of trust in one's abilities is involved, and more generally, motility as exchangeable capital is not always a stand-alone element in one's decision to perform movements and/or exchange them.

Definitions and meanings for potential mobilities

Following the review of motility, which presented also several elements of potential mobilities, we turn now to an explicit discussion of potential mobilities. This discussion will attempt to crystallize and to expand on potential mobilities inclusive of components of motility. We believe that potential mobilities are relevant to two spheres of mobility: individual potential mobilities and societal ones. By individual potential mobility we refer to the capacity of persons to acquire and/ or hire mobility media and to operate them (involving access and competence), the availability of these mobility media prior to any specific movement, and to individuals' decisions (or appropriations) to make use of any of these media for any given planned movement. By societal potential mobility we refer to the cumulative choice of mobility media of people within a given territory and/or at any given time. Let's expand below on these two types of potential mobility.

In the definition of individual potential mobility proposed here we exclude practiced mobility/movements, though we will mention later the potential impact of practiced mobility on future potential ones. We further exclude the supposed autonomous potential mobility of goods and information. On the other hand, we believe that a person's potential to move at any given time to any specific destination, notably through corporeal mobility, constitutes a form of capital which may be exchanged for other people's mobility.

It is of interest to examine the components of potential mobilities at times of technology-based mobilities. One may assume that the basic human motivations for mobility (the 'push' and 'pull' effects) outlined before and in Chapter 2 have not changed in the contemporary technology-based mobility age, with the possible exception of increased levels of curiosity, given the enormous and instant satiation of curiosity made possible through the vast information resources available on the Web.

The access element of potential mobility has changed extensively through the growing access of mobility actors to mobility technologies, particularly virtual ones. The mobile phone has turned out to be the most widely diffused communications device, with a global penetration rate of 87 percent in 2011. This high percentage actually represents a much wider availability of mobile phones, since in developing countries people rent their mobile phones to others as a commercial service. Furthermore, as of 2002 the percentage population worldwide owning a mobile line has been higher than that having a fixed line (ITU 2010a; ITU 2011). The global population percentage of Internet adoption lags behind both fixed-line and mobile telephones standing at 25.9 percent by the end of 2009, but it should be noted that Internet use requires literacy, so that personal competence restricts access to it (ITU 2010a). There are, therefore, three technologies for virtual mobility with wide global distribution: the mobile and fixed telephones and the Internet with various combinations among them (see Chapter 7).

Access to private vehicles as a medium for corporeal personal mobility, is still globally lower than access to virtual personal mobility media, given its higher prices for both purchase and use. The global rate for private vehicle ownership stood at 18.3 percent of the population in 2007 (The World Bank 2010). However, the rate of private vehicle ownership in developed countries is obviously much higher peaking with some 82–90 percent in the US in 2007 (The World Bank 2010). Besides private means of transportation there are, obviously, various public transportation systems available for physical mobility, something which is unavailable for virtual mobility, other than the non-instant postal service (see Chapter 6).

Side-by-side with the widening access to mobility technologies, the global economy has growingly been based on the *space of flows* defined by Castells (2000) as "the material organization of time-sharing social practices that work through flows" (p. 442), where 'flows' include all possible ones, except for people: capital, information, technology, organizational interaction, images, sounds, and symbols. These growing flows have forced people to cope on a daily basis with their own potential mobility or flows, since the need to be mobile, at least virtually, has become incorporated in people's lives. The space of flows constitutes just one expression of a society which has become mobility-based (see e.g. Urry 2007).

Societal potential mobility refers, as mentioned above, to the cumulative choice of mobility media within a given territory and/or at any given time. It focuses, therefore, on the potential mobility *of* a population defined by a specific territory or at a specific moment in time, whereas the potential mobility of individuals focuses more on their potential *for* mobility, as expressed by their appropriations in light of access, competence and social-cultural factors. Societal potential mobility has received little attention so far (see brief comments in Kesselring and Vogl 2008, Canzler *et al.* 2008, Hannam *et al.* 2006). Urry's (2007) suggestion for network capital as a societal asset, mentioned before, may be extended to include also potential network capital. Based on the works of Sorokin (1927) and the Chicago School (e.g. Lannoy 2003), Kaufmann (2009) proposed a wide view of mobility as *social change*, including both, movement and mobility, as well as motility, as intention and ability for change.

A special urban example for societal potential mobility is Manhattan. One may assume that the population of Manhattan could have reached at least the US national car ownership rate of 82 percent in terms of its *potential* access and competence. However, the car ownership rate there stands at only 23 percent, and if Manhattanites had materialized their ability for car ownership then the additional 4.5 million cars would have required a parking lot the size of the whole of Manhattan Island (CEOS for Cities 2010). The population of Manhattan is no less mobile than other American urban populations; it just avoids car ownership because of the expensive parking costs, heavy traffic congestion, and the availability of alternative abundant public transportation media. Thus, some 72 percent of Manhattanites used public transportation for commuting in 2001 and only 18 percent preferred to drive, as compared to the nationwide US values of 88 percent and 5 percent respectively! (RITA 2010). By the same token, 2.4 million out of the 4.3 million commuters into Manhattan in 2008, or some 56 percent of them, opted for public transportation rather than using their own cars, and one may safely assume that almost all of them own private cars (US Census Bureau 2010a). From a potential mobility perspective, the potential mobility of most Manhattanites excludes access to private cars, and the appropriation regarding commuting of more than half of the commuters to Manhattan excludes the use of private cars. Both cumulative decisions, amounting to societal potential mobilities, make it possible for Manhattan to constitute a well-functioning global business center.

The same may apply also for societal virtual mobility. Mother's Day is considered the day with the highest volume of telephone calls throughout the year in the US and in various other countries. In the US this traffic volume is 8 percent higher than on New Year, the second busiest day for telephone traffic (Reuters 2010). In the past, when only one telecommunications medium for calling existed, the fixed-line phone, the day was characterized by traffic jams, causing troubles and delays to callers. The wide diffusion of both mobile phones and VoIP (Voice over Internet Protocol) calling has spread the calls among these three calling avenues, thus reducing traffic congestion. From a potential mobility perspective, for each caller the availability of three calling channels represented both wider access and, thus, more complex appropriations, whereas for all callers it meant easier and better contact.

Active and passive potential mobilities

Flamm and Kaufmann (2006) accentuated the difference between motility and accessibility by stating:

> Motility differs from accessibility by focusing on the logic of an actor's actions, in particular the reasons behind the choice of tools and localizations, without

being concerned with an action's maximum utility. In this sense, motility concentrates more than accessibility on how an actor builds his/her relationship with space and less on the possibilities offered by a given territory (p. 169).

In this statement motility is presented as an active potential mobility in the sense that an actor makes choice among several mobility possibilities, whereas accessibility is presented as a parameter related to territories and not to people.

One may, though, extend the logic of accessibility into the logic of potential mobilities, by treating accessibility as providing passive potential mobility. In other words, the accessibility of a person to other mobile actors depends on her/his very availability for contact by these other actors. Consider, for example, a person who does not own a private car and relies for mobility solely on public transportation as well as on walking to the nearest bus/metro station. Though such a person has no access to a private vehicle, usually he/she is still accessible to other mobile actors who own a car and can reach her/his home by their cars. Thus, the lack of some kind of active potential mobility does not avoid having a passive one of the same kind via accessibility to others who have such an active potential mobility.

When it comes to virtual mobility, accessibility or passive potential mobility is different. If somebody does not own a certain medium (i.e. fixed or mobile telephones, or the Internet), then he/she has no access or active potential mobility through that medium, and simultaneously she/he is not accessible through that specific medium to others who have an access to this medium. In other words this person has no passive virtual potential mobility through that medium. In a more extreme case, if one does not own any of the media for virtual mobility and does not rent their services, then this person has no virtual active and passive potential mobilities at all.

Potential mobilities and practiced ones

By their very nature, potential mobilities do not constitute 'stand-alone' entities. Rather they are a preparatory step towards practiced mobility and specific movements. The appropriation phase of potential mobilities is continuously redesigned by the growing availability and access to mobility media which characterize the contemporary age. This wider access implies increased flexibility in practiced mobilities as well as better fitting of media to specific mobility requirements.

The relationship between potential and practiced mobilities is a two-way road, with potential mobilities paving the road for practiced ones, and *vice versa*: practiced mobilities shaping future potential ones. Thus, experience gained in practiced mobility may influence potential mobility. Such an experience may impact access by convincing a mobile actor to buy or to sell any mobility medium whether for corporeal or for virtual mobilities. Less dramatically, practical experience may lead to better future appropriations as to which medium should be used for any given movements. Furthermore, the practical use of new media, such as smartphones or *Facebook* may imply new competences, and thus differing appropriations.

A model for potential mobilities

Based on our discussions so far we can suggest a simple model for potential mobilities (Figure 4.1). It is assumed that potential mobilities, like their resulting practiced ones, stem from basic mobility needs, which are three: 'push' effects (consisting of locomotion, proximity, and curiosity); 'pull' effects (including people, places, events, and information); and daily needs (such as commuting and shopping), the latter serving as mobility triggers. Potential mobilities *per se* are conditioned by three factors: access to mobility media by way of their adoption by mobile actors and their daily availability to them; competences or skills for the use of mobility media; and social and cultural contexts for any potential mobile actor, such as cultural restrictions, family relations, etc. Mobility needs, motives, and aspirations, along with access to mobility media, competences for their use, and social and cultural contexts, are all weighed by mobility actors in the appropriation phase, bringing about choices of mobility modes and media and decisions on their immediate uses.

A mobility actor may opt in principle for one of three mobility options. One such option is avoiding any kind of mobility, neither an active one of reaching out physically or virtually, nor a passive one which have permitting other mobility agents to access and reach them. A second option of mobility is choosing an avoidance of active potential and practiced mobilities but permitting passive ones, by being accessible to others who would move to the passive mobile actor. The third, and more frequent option, is potential active mobility, implying a choice among various mobility media at large, and for specific movements in particular. Another, and fourth, mode of potential active mobility would be a mobility actor's decision to relate to mobility as capital and have it exchanged with other people's mobility.

Practiced mobilities, whether through active movements by mobility actors, whether by other people moving *to* them (passive mobility) or by other people moving *for* them (exchanged mobility), contribute to busier societal landscapes of transportation, characterized by public and private terrestrial transportation media, as far as corporeal mobility is concerned, and to increased telecommunications traffic as far as virtual mobility is concerned. Furthermore, and back to individual mobility actors, each of their specific movements does not only move themselves or others moving to them or for them, but it also adds to their own mobility experience. This experience may become of importance for future potential mobility choices and decisions.

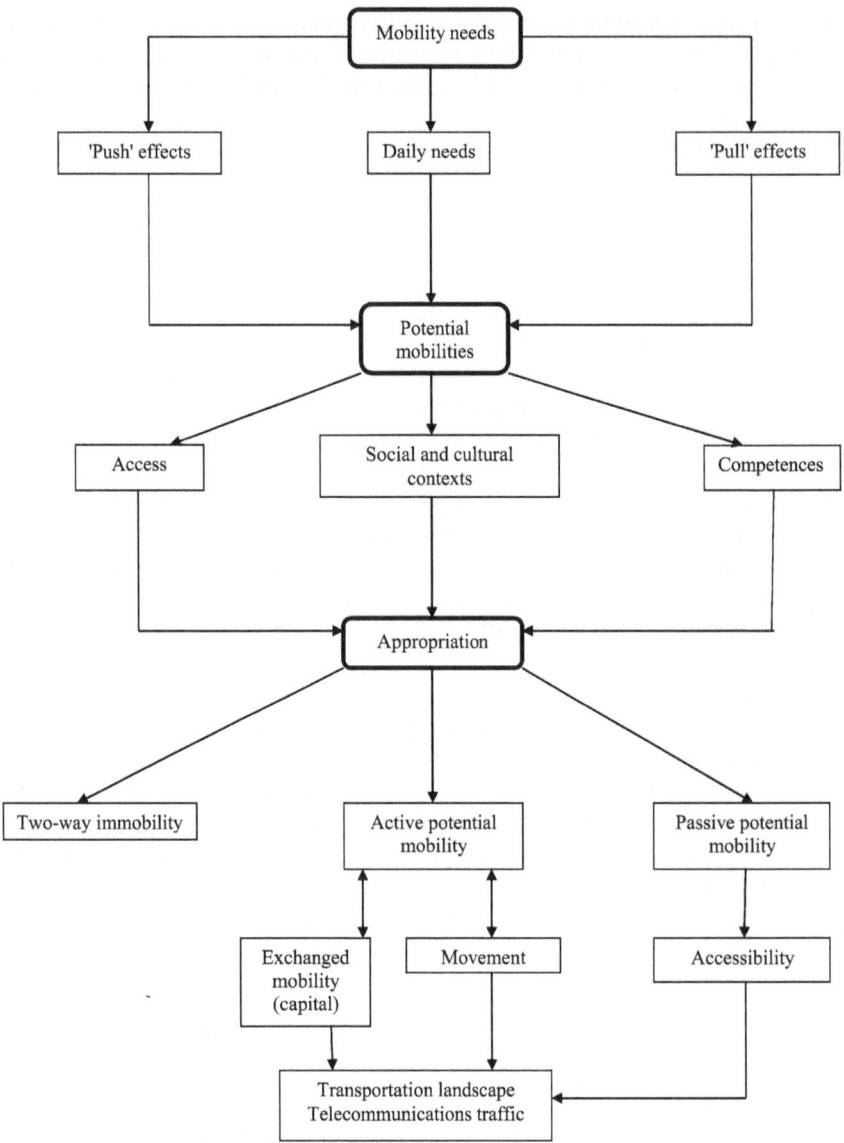

Figure 4.1 A model for potential mobilities

Source: Kellerman 2012.

Conclusion

The objective of this chapter was to develop and expand on the concept of potential mobilities in light of and beyond motility. The discussions and the proposed model are conceptual in nature, and as such they cannot be tested directly in an empirical study, such as Kaufmann's (2002) empirical study on motility. However, the elements for potential mobilities proposed here may serve as building blocks for future studies attempting to assess appropriation processes, or for studies attempting to evaluate potential mobilities in light of practiced ones. Another dimension which requires future work is societal potential mobilities, or potential mobilities as capital and resource of cumulative value for social groups.

The discussions in the foregoing sections attempted to contribute to Kaufmann's (2009) call for "a comprehensive approach to mobility that explores [these] logics of action" (p. 53). This perspective of action is important since a lacuna seems to exist regarding the understanding of potential human action at large, not just regarding mobility. Exceptions to this are the geographic discussions relating to action space and activity space (Golledge and Stimson 1997, see Chapter 1). Potential mobility may serve as a building block for the study of potential human action at large, beyond the specific action of mobility, by offering a general framework of five major phases for human action, leading from the development of potential actions into practiced ones: needs and triggers; accumulation (access, competences, social and cultural contexts); choice (appropriation); practice (mobility in our case); and new cycles of potential and actual action, based also on experiences gained in previous cycles of appropriation and practice. Obviously, such five-phase structures for the interpretation of any action are not constant, as change in each of these phases has to be considered, as Kaufmann (2009) proposed for motility. The contemporary social and economic worlds which accentuate intensified and accelerating action at large, call for a clear understanding of the paths leading all the way from intentions to act through action potentials to practiced action.

Chapter 5
Mobility or Mobilities?

In our discussions so far we used the terms 'mobility' and 'mobilities' as interchangeable with each other. In this chapter we will attempt to assess these two notions in light of various mobility types (terrestrial, virtual and aerial), in order to see whether mobility could be considered as one single concept or whether the differences among the various mobility types call for viewing them as separate mobility categories. This discussion of differences and similarities among mobility types will, by its very nature, highlight numerous functionality dimensions of daily spatial mobilities. Thus, the chapter will add the functionality dimension to the roots and contexts for potential and practiced daily spatial mobilities which we presented so far. Furthermore, in some way this chapter will pave the road for the second part of the book in which we will examine separately terrestrial, virtual and aerial mobilities.

Potentially, there are actually four categories of mobility over space at large: terrestrial, maritime, aerial, and virtual. The vast developments and enhancements of technology-based aerial and terrestrial mobility have turned long-established maritime mobility, using boats for functional mobility, into mostly pleasure-oriented cruises. Even before the introduction of technology-based mobilities, maritime mobility by boats was used for long distance travel but could not be used for short duration daily mobility given the slow speeds of sail and rowing boats (with exceptions such as Venice and Amsterdam). By the same token, the crossing of straights and canals was left mainly to ferries, which may be viewed as complementary and extension means for terrestrial mobility, frequently permitting the joint carrying of passengers and their cars. These trends were well reflected in the 2006 percentage modal split of passengers in EU-27 countries: 72.7 passenger cars; 8.6 air; 8.3 bus and coach; 6.1 railway; 2.4 motorcycles; 1.3 tram and metro, and merely 0.6 percent ferries (Larsson and Götaland 2009). We will, therefore, exclude maritime mobility from the following elaborations.

The need to comparatively and jointly examine and assess the various types of mobility has been recently expressed regarding aeromobilities. In his introduction to a reader on aeromobilities, Cwerner (2009) claimed that "aeromobilities research needs to link up with the wider field of mobilities studies…aeromobilities are related in several different and distinctive ways to other forms of mobility" (p. 10). This chapter attempts to do just this, albeit within the wider context of the three leading mobility types. It attempts to highlight the distinctive features of each mobility type, as well as convergences among them, leading to a discussion which will assess whether these three mobility types constitute just categories of a single entity of mobility, or whether they, alternatively, constitute three distinct mobility entities. We will attempt to show that the three categories constitute jointly a single mobility entity, with all categories sharing a basic model of mobility cycles: 'push' and 'pull' effects of mobility motivation, followed by the process chain of IT→origin→route→control→destination (Kellerman 2011).

This exercise is of significance because of the tendency to refer to 'mobility' in a rather general sense. It is questionable whether this general entity makes any conceptual or practical sense, from the perspective of the three mobility types. Perhaps we should always have to add a proper category to the rather general term 'mobility', i.e. terrestrial, virtual, maritime, or aerial, because of the distinct meanings and operations of each of these mobility types. If the latter option is to be preferred, then the general entity should rather be 'mobilities', in the plural form. Usually, examinations of concepts and phenomena lead towards their partition or classification into categories for the sake of their framing, organization or arrangement (Demarco 2004). Thus, "any research field needs to classify entities and phenomena into categories" (Aurnague *et al.* 2007: 1). Here our attempt regarding mobility/mobilities will be the opposite, assessing whether the very existence of three specific types of mobility (terrestrial, virtual and areal) should dictate viewing them as separate entities of mobility, or whether they should be viewed simply as categories of a single more general entity of mobility.

In the following sections we will, first, review the topic and set it within a wider mobility context. This will be followed by an elaboration on the features of each of the three mobility types: terrestrial, virtual and aerial, through their comparison with each other, highlighting mainly the numerous differences among them. The next section will attempt to present an alternative perspective through the examination of convergences among the three types of mobility. These two discussions will lead us to an assessment of whether we may refer to mobility as a general single entity or whether the three types of mobility imply also the constitution of three distinct mobility entities or categories. We will propose to adopt the option of mobility as a single entity, and will exemplify this approach/ model through comparisons of terrestrial public and personal mobilities, as well as through aerial and virtual mobility modes. We will then conclude with brief discussions of potential implications of viewing mobility as a single entity.

Mobility and/or mobilities

The following review and proposed conceptual structure are aimed at two ends. First, to highlight some literature on the terminology of the person form used for the social conception of motion: mobility or mobilities. Second, to propose a conceptual structure which will put the mobility/mobilities question within the wider contexts of mobility concepts and theory. This proposed conceptual structure will be based on the viewing of mobility/mobilities as nesting within the double sense of the term 'mobility', social and spatial, with spatial mobility being our focus here. It will further be based on the several motivations for daily spatial mobility

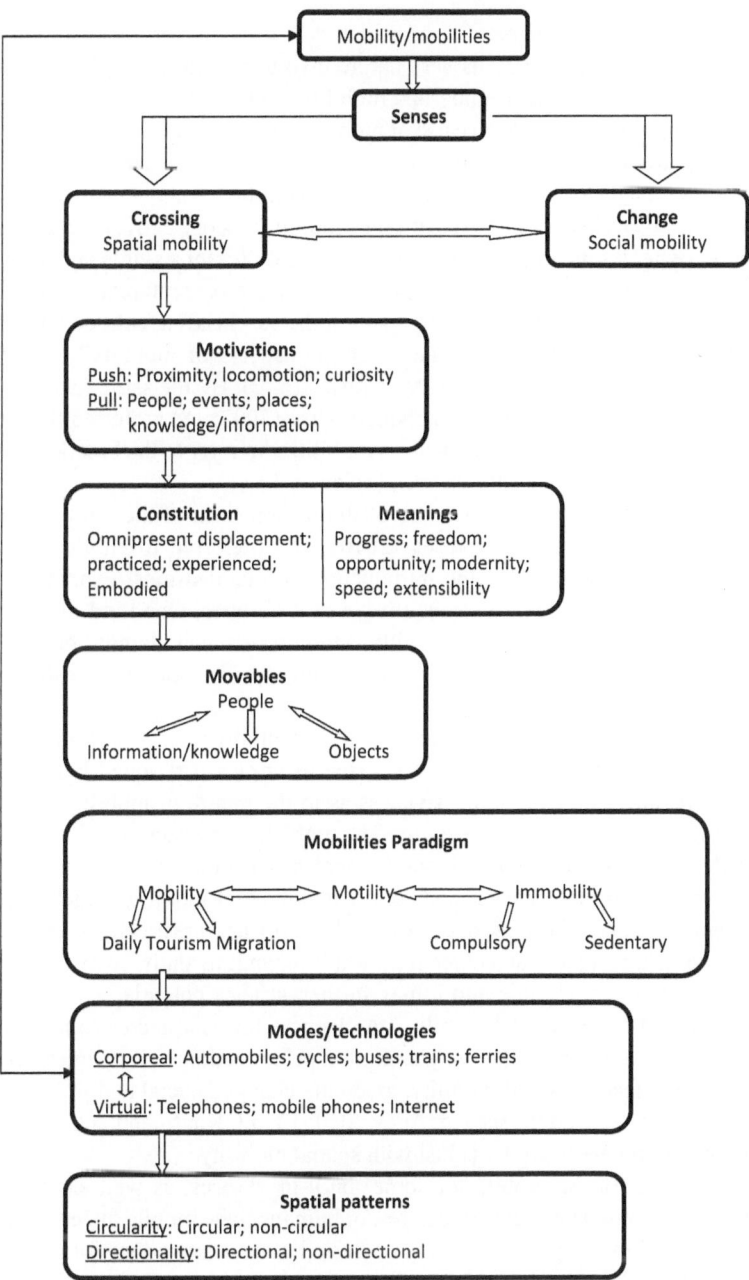

Figure 5.1 The context of mobility/mobilities

Source: Kellerman 2011.

which we discussed in Chapter 2, and which determine its very constitution and meanings. Spatial mobility by its very nature involves various movables (people, information and goods) and numerous mobility options (terrestrial, virtual and aerial), all implying an integrative use of mobility technologies and the production of spatial patterns. All these components of daily mobilities are unified by the mobilities paradigm which we will discuss later (Figure 5.1).

Urry's (2007) book-length exposition of mobility theory carries a plural title, *Mobilities*, whereas its objective was to develop and present a single and unifying structure for mobilities: the *mobilities paradigm*. This mixed reference to mobility in both singular and plural is exposed again in the sub-heading entitled "different mobilities", which is said to outline "multiple aspects of mobility" and "four main senses of the term 'mobile' or 'mobility'" (p. 7). By the same token, when presenting his unifying mobilities paradigm, Urry (2007: 18) preferred the plural form of the term mobility: "I use the term mobilities to refer to the broader project of establishing a movement-driven social science."

An opposite, seemingly mixed, use of the person form of the term mobility/ mobilities is a recent book by Adey (2010) who preferred to entitle it in the singular: *Mobility*, whereas in the introduction he claimed most explicitly: "To speak of mobility is in fact to speak always of mobilities. One kind of mobility seems to always *involve* another mobility. Mobility is never singular but always plural. It is never one but necessarily many. In other words, mobility is really about being mobile with" (p. 18).

Using the term mobilities in the plural may potentially refer to multiple aspects of mobility as a singular term. However, the examples mentioned before may, alternatively, attest to some indecisiveness as to the nature of mobility, possibly constituting a single or a multiple entity. The problem of person form for mobility is rooted in the most basic sense of mobility which is a double one: 'vertical' social mobility which implies change, and 'horizontal' spatial mobility which implies movement in space. These two senses of the term are practically interrelated, since moving up the social ladder may imply changing daily spatial mobility patterns, notably commuting, and it may further, at least potentially, bring about or be attributed to residential mobility, to business tourism and/or to migration. This may also be true the other way around: a person's domestic or international migration may lead to social mobility, given the changed social and professional milieus in the new location (see e.g. Adey 2010: 13, 37, Kaufmann 2002, Chapter 1). Our attention here however is just with spatial mobility.

Humans need to be mobile for some 'built-in' desires, as well as for some external attractions. The needs, or desires, of humans to be mobile in general have been elaborated in Chapter 2, focusing on basic human motivations for proximity (to other people), locomotion, and curiosity (for information/knowledge). These motivations we termed as 'push effects', since they imply a desire to reach out or move away from a certain location, physically or virtually. As we have seen, the other side of the coin of human needs to move is the attractions provided by any destination, and which may potentially consist of people, events, places, and information/knowledge (Urry 2002). These attractions may be viewed as 'pull effects', drawing people to them, and, thus, bringing about their movement in order to reach them.

As we have seen in Chapter 1, spatial mobility, stemming from 'push and pull' motivations, constitutes foremost a constant, omnipresent displacement of people, objects and information, differing by their flexibilities. Spatial mobility in the form of migration has been shown by Cresswell (2006) to constitute a major cultural experience. Our focus here is rather on daily spatial mobility, consisting of commuting, browsing, communicating/networking, shopping, social and community visits, and business trips, all of which integrate physical and virtual mobility.

Human corporeal and virtual mobility can be conceptualized through its various forms as well as through its durations of materialization. These ideas were advanced by Urry (2007) under a single conceptually integrative umbrella, the *mobilities paradigm*, which "emphasizes the complex assemblage between these different mobilities that may make and contingently maintain social connections across varied and multiple distances" (p. 48). Thus, the paradigm includes three major mobility conditions. The first is experienced mobilities, consisting of daily mobility (commuting as well as social travel, e-mailing, phone calling, etc.), less frequent business and pleasure tourism mobility, and long-term, one-way migrations. The second mobility condition in the paradigm is *motility*, as suggested by Kaufmann (2002: e.g. 37) for potential mobility, or various social and economic human capacities for integrated social and spatial mobilities. The third mobility condition is rather immobility, which may reflect two realities: a compulsory one, in order to make it possible for other people to move (see e.g. Boltanski and Chiapello 2007), when "the social and built environment empowers some to be more mobile *at the expense of others*" (Sheller 2008: 29), and voluntary sedentarism, when people prefer not to be mobile.

People may use various types, modes or technologies for their mobility, and this choice is a basic assumption for the discussions in this is chapter. They may wish to move corporeally, using automobiles or cycles for personal mobility, or they may prefer trains, buses, taxis, ferries and planes for public transportation, permitting terrestrial, aerial and maritime movements. They may further wish to use virtual mobility modes, mainly fixed telephones, mobile telephones and the Internet, separately or jointly with corporeal mobility, thus enjoying written, audio and/or visual communications. This variety of mobility technologies, providing numerous options for movement and displacement, raises the question, which is at the core of our discussion in this chapter, whether, from a perspective of mobility types, all these options comprise a single mobility entity, as may be implied from the discussion so far, notably given the unifying mobilities paradigm, or whether they constitute three separate mobility entities.

The spatial patterns of movements for the meeting of people, events, places, or information, may reflect various mobility conditions and numerous mobility modes/technologies, as well as distinctions by travel frequencies and destinations. Thus, repetitive and frequent movements between the same origins and destinations, present *circulation* (e.g. commuting to work or to school; daily or weekly shopping; regular visits of family and friends; see Chapter 1) (Amin and Thrift 2002: 81-3), whereas movements from a particular geographical origin to one or many unknown destinations, constitutes *non-directional mobility* (e.g. web surfing by Internet users; see Chapter 1) (see Bonss and Kesselring 2004). Other combinations between circularity and directionality are also possible. For instance, Internet surfing may be non-circular in that different sites may be visited at changing times of surfing, but it may also be circular when the same website is used at given times, such as periodic checking of bank accounts. Movements may further be *non-circular directional*, namely non-repetitive but destination-defined. For instance, travel to business meetings, and shopping on special occasions and places.

Features of the three mobility types

We will now turn our attention to mobility through its numerous terrestrial, virtual, and aerial modes. Reaching out corporeally through natural means includes walking, running, speaking, singing, whistling, whispering, shouting, body language, as well as bodily contacts. The connectivity of humans involves, therefore, both reach and access. People need simultaneously to reach each other, as well as to reach events, places and information. As we have seen in the previous chapter, individuals further need to have access and to be accessed in order to reach these elements of attraction. Our focus here will not be on natural mobility, but rather on the technology-generated form, permitting movement in space terrestrially, virtually, or aerially, through the use of mobility technologies. Table 5.1 presents details for the various features of each of the three technology-based mobility types, divided into features of mobility systems/media and features of people moving through the use of mobility technologies available for each of these mobility types.

The three mobility types seem to widely differ from each other, beginning with the most basic feature of the physical size of mobility media and their spatial infrastructure context. Virtual mobility means (PCs, laptops, telephones and mobile phones) are small in their physical size and in recent decades they have continuously become smaller and smaller, which is also true for the infrastructure which serves them, consisting mainly of wires, exchanges, servers and antennas. On the other hand, terrestrial mobility vehicles and their infrastructures are big and spread all over, including cars, buses and trains, as vehicles, and roads and railways as infrastructures, whereas the vehicles for aerial mobility, planes and helicopters, are even larger in size, but confined mainly to land-extensive airports. Growing demands for all three mobility types have amounted to the construction of three separate and most extensive infrastructure systems. Furthermore, all mobility types have to be widely available, since a basic social requirement for people in contemporary societies is for them to be able to use, actively (through personal mobilities) and/or passively (through public mobilities), the various mobility technologies.

As Table 5.1 demonstrates, the three mobility types differ from each other in their systems, through numerous dimensions: ownership, organization, terminals, vehicles, channels, flows, pace/speed, control, availability, and environmental effects. They further relate differently to human beings *vis-à-vis* the choice, autonomy, operation, preparation, experience, bodily conditions, and socialization which they offer or permit. These wide varieties in the dimensions of mobility stem from the distinctively different character of the movements of the human body when using the various mobility types, on the one hand, and the movability of thoughts and emotions of human beings, on the other, along with big differences in mobility technologies among mobility types. In other words, terrestrial and aerial mobility media permit the transfer of human beings as a whole, including body, thoughts, and emotions, whereas virtual mobility transmits only human thoughts, emotions, side-by-side with actions, such as payments, and information/knowledge. Transmissions via virtual mobility media can still be most powerful, as compared to corporeal movements, notably since the various media for virtual mobility permit instant intensive and extensive movement/interaction sessions, at any time, as compared to bodily movements which can be executed only one at a time, and normally each movement/trip being time consuming until destination is reached. Sometimes people prefer to transmit their non-bodily self through virtual media rather than to move corporeally but still remain internally sealed for meaningful human interactions/dialogs.

Table 5.1 Features of the three mobility types

Feature	Terrestrial mobility	Virtual mobility	Aerial mobility
Systems			
Ownership	Infrastructure mostly governmental; vehicles mostly private or commercial other than most buses and metros	Infrastructure and vehicles mostly private	Infrastructure mostly governmental; vehicles mostly commercial
Organization	Public, commercial, and private	Private except for the postal service	Public only
Terminals	Railway and bus stations; parking lots and spaces	PCs and laptops; fixed-line and mobile phones	Airports
Channels	Roads and rails	Cables and wave spectra	Flight routes
Vehicles	Cars; cycles; buses; trains	None	Planes; helicopters
Flows	Dependent on traffic on road	Dependent on traffic and general and personal infrastructure	Dependent on terminal (airport) load
Pace/speed	Fast (compared to walking and animal riding)	At speed of light (potentially at least)	Very fast (compared to terrestrial mobility)
Control	Governmental traffic controls via IT; no domestic control over passengers	Governmental wave spectra allocations; normally no governmental control over information flows. Seamless traffic control by IT	Full governmental control over the number of flights; flight operations; passengers, and luggage. Manual and IT control
Availability	Private cars always; public transportation by timetables	Always	By restricted timetables
Environmental effects	Pollution; noise; extensive space consumption	Possible radiation	Noise; extensive space consumption

People			
Choice	Between public and personal; often also between train and bus	Between written and voice communications; between computers, telephones, and mobile phones	No choice other than among carriers
Autonomy	Low in public transportation; relatively high in driving	High	None
Operation	Through agents in public transportation; personally in driving	Personal, except for mail service	Through agents only
Preparation	Dependent on destination (local, domestic or international)	None	Extensive for international travel (passport; visa; foreign exchange; insurance; lodging)
Experience	Traffic congestion; impedance for commuting	Co-presence; time-space compression	Time-zone and cultural change
Bodily conditions	Fatigue because of driving	Sometimes eye and hand problems because of abuse of computers and maybe radiation effects of mobile phones	Fatigue for no movement and no sleep; jetlag
Socialization	Limited in cars, less so in public transportation	Extensive by choice	Limited on planes and in terminals

Source: Kellerman 2011.

Convergences among the three mobility types

As we mentioned already, normally studying a phenomenon or concept involves, at one phase or another, attempts to categorize or sub-group the phenomenon or concept under study into distinct smaller entities. Here, we move the other way around, having shown first the differences between the several types of mobility distinguishing them from each other, we attempt now to examine convergences between these same mobility types. Such convergences might bring about viewing the three mobility types jointly as a single possible entity of mobility.

The substantial differences between the three mobility types presented in the previous section are coupled with several convergences, emerging through the vast development and utilization of information technology (IT), which extend beyond the obvious roles of IT for virtual mobility *per se*, as an IT-enabled communications technology. The roles of ITs for mobility at large and their converging effects may be divided, like the differences between the mobility types, into their significance for mobility systems and their significance for moving people.

From the perspective of mobility systems, ITs have become the controlling technologies for virtually all mobility types and in most of their dimensions. Access to mobility is controlled by IT, whether it is through the purchase of train, bus, or airline tickets, or whether it is through car ignition, and computer or telephone operations. Movement itself is controlled by IT in at least two ways: controlling the very movement of mobility vehicles which serve specific moving persons (e.g. through automatic steering systems for cars, buses, and planes), and, simultaneously, also controlling all the moving vehicles and information transmissions at any given time, through controlling facilities or devices such as traffic lights, control towers or routers. Beyond these automatic controls, workers in traffic control functions use communications media for communicating with each other in order to assure the proper control over the movement of cars, trains and planes. At the end of each movement i.e. the very stopping of any vehicle, may too be controlled by IT. Thus, IT is there in more than merely service-oriented software as proposed, for example, by Dodge and Kitchin (2004) and Dodge *et al.* (2009). In other words, the corporeal travel of people, terrestrially and aerially, is coupled with vast transmissions of information between machines, as well as between human controllers, probably amounting to a significant portion of the velocity of transmissions of information among humans involved in virtual mobility *per se*.

From the perspective of users, Cwerner (2009: 10) noted for aeromobilities that they "are related in several different and distinctive ways to other forms of mobility, sometimes as rings in a chain of intermodal transportation systems." Thus, passengers may reserve flights, purchase and receive tickets and boarding passes through the Internet and even through mobile phones, then move by terrestrial mobility modes to airport terminals, and following the landing of the flight may again use terrestrial and sometimes virtual mobility media.

Mobility types are not just interconnected as chain rings but they may be used simultaneously, as well. Thus, it has become common practice to use mobile phones while driving or while riding any vehicle of terrestrial mobility, and the Internet can be used while travelling on public mobility vehicles. The Internet and mobile phones have been shown to be effective political media as well, for instance for the organization and coordination of street protests, such as in recent events in Iran, North African and Middle Eastern countries. However, the use of virtual mobility media on planes is still more restricted. Some planes are equipped with rather expensive satellite telephones, and experiments have been performed with possible uses of mobile phones and the Internet on board. In addition, and in what has become obvious, virtual mobility media permit communications by passengers from their destinations calling back to their places of origin at varying costs, depending on location and chosen media. Another almost simultaneous and integrated potential option of linking terrestrial and aerial mobilities, albeit still not available for civilians, is carrying cars with their drivers on planes, as is common on ferries.

Mobility categories or mobility entities?

The current state of the art of mobility and the convergence of mobility modes may permit one to claim that the three mobility types constitute categories of a single mobility entity for both moving people and for mobility systems rather than their constitution of three distinct mobility entities. This claim is based on two models. We will first examine a simple joint model for people moving terrestrially, virtually, and aerially, followed by a similar exercise for the mobility systems of these mobility types.

The first model shows people's mobility through any chosen mobility type and media (Figure 5.2, see also Kellerman 2006a: 46-9). The model assumes common structures of mobility cycles from origins to destinations, no matter which mobility media are used. The mobility practice of human beings has both been argued for, and has shown evidence, that people wish to perform corporeal movements as a common daily practice, even if some movements can now be accomplished virtually, so that the time saved through the use of virtual mobility may be used for new corporeal movements. Thus, it was argued that the total amount of time devoted to the use of all mobility media has remained quite constant (see e.g. Hupkes 1982, Mokhtarian and Chen 2004, Kellerman 2006a).

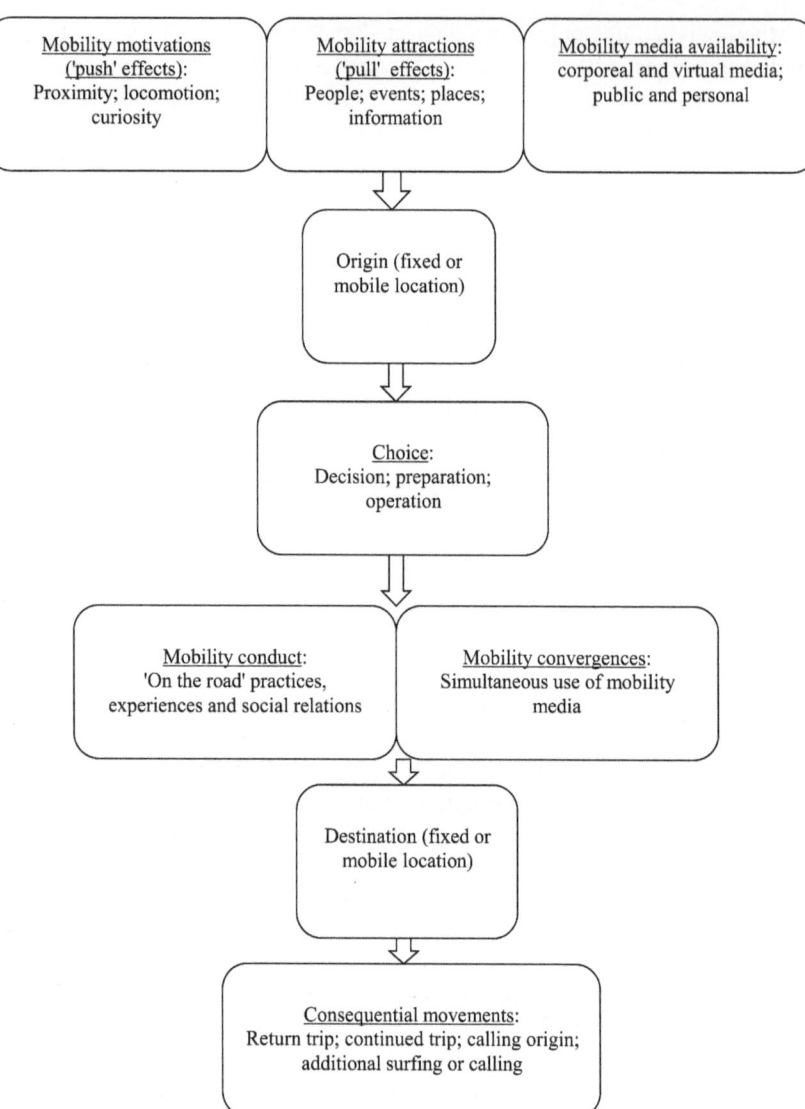

Figure 5.2 Mobility cycles for people

Source: Kellerman 2011.

The starting points of the model for moving people are the 'push' and 'pull' effects for mobility, which we discussed in Chapter 2. They include, as push effects, the needs of humans to be mobile in general, and to reach out from their points of origins, reflecting basic needs for proximity, locomotion, and curiosity. They further include, as pull effects, attraction to people, events, places, and information/knowledge at any destinations, calling for moving to them by any mobility type and media. The spatial pattern for reaching any destination may involve some combination of circularity and directionality, as mentioned already in a previous section.

Decisions on particular movements may sometimes be automatically geared towards physical mobility, whether through walking, public transportation, or through the use of private vehicles. Such automatic preference for corporeal mobility may occur either for daily routine tasks, such as commuting, or when it is clear that physical proximity should be the first choice for interaction. Another choice can be made among the numerous virtual media for movement, whether it is the telephone, the fax or a complex Internet session, beginning with an information search over the Web, and followed by e-mailing or other forms of messaging. Licoppe (2004) described the variety of communications media, as ranging from delayed response (e.g. SMS or e-mail), through co-presence in time only (telephone) to time-space co-presence (face-to-face virtual and corporeal meetings).

Once on the move, whether physically and/or virtually, moving individuals are engaged in specific behaviors, which involve practices, experiences and social relations, each reflecting opportunities *vis-à-vis* the chosen mobility media, and obviously reflecting the individual character of the moving individuals. The elements of on the road behavior are mentioned here only briefly as they were discussed in detail elsewhere (see Kellerman 2006a). Certain practices typify driving personal mobility vehicles (cars and cycles), and these include routing, maneuvering under circumstances of road congestion, and visibility. On-the-road experiences include ergonomics and hybrids between drivers and their computerized 'intelligent' vehicles, driver contacts with the environment, and road directionality, whereas the third and last category of on-the-road behavior which relates to social relations may consist of three major aspects: filtering of contacts with other drivers, potential gallantry and/or rage when relating to other road users.

Persons who are moving may simultaneously engage in virtual mobility as well, through the use of mobile phones, or laptops, for both telephone calls and accessing the Internet. The same may apply also for the use of such communications means while spending time at destinations when on physical travel. Reaching another place may call for continued travel as well, and similarly reaching a certain website through surfing may bring about additional surfing to other websites.

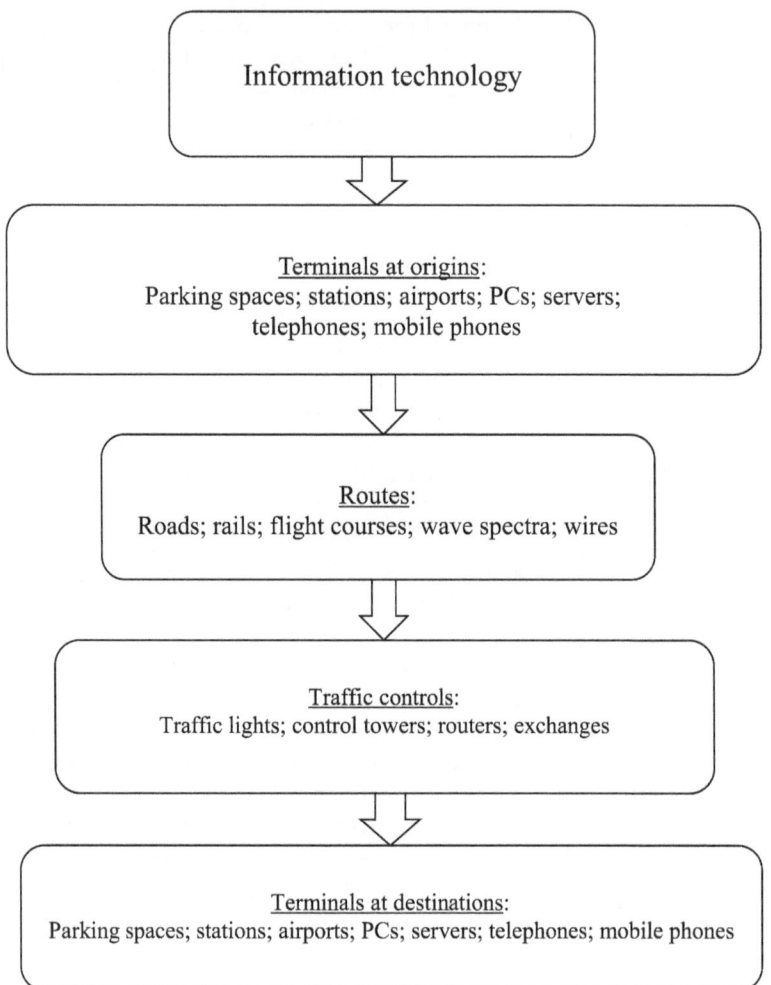

Figure 5.3 Mobility cycles for mobility systems
Source: Kellerman 2011.

    The second model recognizes the intrinsic role of IT in the mobility systems
of the three mobility types as a single mobility entity rather than their being
three distinct mobility systems (Figure 5.3). This model is based on information
technology as a shared technology for the operations of mobility systems for all
three mobility types, as well as for the shared existence of terminals, routes, and
traffic controls in all three mobility types. Kakihara and Sørensen (2002: 131, see
also Urry 2007: 53-4) termed ITs as "essential 'blood vessels'" for contemporary
mobility systems. The basic structure of this model is, thus, similar to the one
proposed before for moving persons. The same technology, information technology,

is used in all mobility modes, providing a common denominator similarly to the basic mobility motivations for humans bringing about mobilities through all three mobility types. IT is the common denominator for the operations of all terminals of origins, as well as for all terminals of destinations at the conclusion of a mobility cycle. Another similar domain is 'on the road conduct', which for humans implies practices, experiences and social relations, whereas for mobility systems it analogically implies keeping the proper track for traffic through mobility systems (i.e. proper lanes, tracks, courses and wave spectra respectively for all modes), governed and controlled by IT.

We can analogically compare the two models, for people and systems, by arguing that the basic needs of humans to move and their attraction to various elements located elsewhere constitute the primitive dimension of human mobility, as compared to information technology which currently is shared and used by all mobility modes and analogically serves as the 'primitive', or basic, element for technologically-driven mobility systems. Information technology operates and governs all mobility phases, including terminals, vehicles and routes. Furthermore, contemporary sensor-laden information technology has the potential for the provision of extensive hybrids and coordination among mobility technologies and vehicles (e.g. between trains and cars), as well as among moving persons and the mobility technologies at their disposal (see e.g. Thrift 2004a). GPS for driving or navigating through a mobility route, and LBS (Location Based Services) for urban service provision or destination fixing, are examples of such technologies which are currently in a process of mass adoption, and are both available through the most mobile mobility device, the mobile phone.

Information technology is applied to diversified tasks in mobility cycles, such as communications, synchronization, controlling, routing, and management. The whole transportation industry has been transformed from low-tech to high-tech, through the incorporation of ITs into these tasks. This process began with planes, moving through cars and airports to buses, trains and stations. As such IT is able to take care of the basic needs of individuals for mobility in the widest possible sense: enormous numbers of people moving simultaneously through varied media, at different speeds, and attempting to reach destinations at differing geographical ranges. It is this role and ability of information technology which has made it possible for mobility to become such a central dimension of contemporary life. The central role of IT in the facilitation, coordination and controlling of mobilities at large which further permits a view of mobility as a single entity consisting of three mobility categories/types rather than as three separate entities.

Terminals are there in order to serve moving people and their vehicles, and terminals are there also for the movement of information which travels without people's bodies and without vehicles. Terminals constitute both the points of origin and the points of destination for moving people, and these points are always there, whether mobility cycles are instant, at the speed of light (as in virtual mobility), whether speed is relatively fast but still time-consuming (as in aerial mobility), or whether speed is much lower and the duration of movement flexible (as in terrestrial mobility).

By the same token, it is not only for people to have to obey certain rules and regulations when on the move, but it is also for mobility vehicles operated by individuals to make sure that they use the proper paths of movement in space and that a proper timing was fixed for each movement through traffic control devices. However, it is the non-automatic choice for mobility media available to humans, or a basic autonomy of humans, which constitutes the clearest dividing line between moving people and mobility systems. This choice has been customarily restricted to the choice made by moving persons between corporeal and virtual mobility, as well as to their choice between personal and public mobility media. However, IT has permitted much more flexibility of choice among virtual mobility media, through combinations and integrations among several virtual mobility media. Thus, one can send a fax from a fax machine and have it received through the Internet and *vice versa*, or one can make phone calls from a computer through VoIP (Voice over Internet Protocol), or send e-mails from a mobile phone. In other mobility types this flexibility is usually impossible: a plane cannot land in a railway station, and for terrestrial mobility a car does not ride on railway tracks.

We will move now to two comparisons between two media in each mobility type, attempting to show that different mobility modes or types can lie along a continuum or they may be integrated with each other, making all mobility modes and types belong to a wide single entity of mobility. The first example will compare public and personal terrestrial mobilities, which are two modes within the terrestrial mobility type, showing that they present some continuum of personal autonomy (see Chapter 3). The second example will compare two mobility modes belonging to two different mobility types: virtual personal mobility via mobile phones as compared to aerial public mobility via airlines. Virtual mobility will be shown to be highly integrated with aerial mobility. Both examples present simple daily patterns but despite or because of their simplicity they may highlight our argument for mobility as a single entity.

Example 1: Public *versus* personal terrestrial mobilities

By public mobilities we refer to public transportation, mainly buses and trains, whereas by personal mobilities we refer to the driving of either automobiles or cycling vehicles. As we have seen already in Chapter 3, public and personal mobilities differ from each other mainly in the degree of personal autonomy which they offer to moving persons. Driving a car implies autonomous mobility in that persons direct themselves corporeally from origins to destinations. Being autonomous in the sense of car driving still implies user reliance on road infrastructures, as well as on complex traffic control systems which facilitate the autonomous driving of cars and cycles. Thus, as we mentioned already, it is obvious that one cannot move from one point in space to another one using a straight and shortest possible route, free of any obstacles; even expressways have their own spatial contours. Driving is rule-bound, and it involves, among other restrictions, no-entry and one-way streets, and it further involves traffic jams. Time friction and space and distance frictions in personal mobility through car driving, might be either higher or lower than those of public transportation, but personal mobility moves one from an exact origin/terminal to an exact destination/ terminal as compared to bus/railway stations which serve wider areas using fixed stops and stations.

Riding buses and trains constitutes forms of public mobilities, without any personal directing or driving of the mobility vehicles. However, even these public mobility services still imply some personal autonomy in the sense of fixing the timing of each ride by the moving persons using system schedules if relevant, and by these persons' choices of the preferred mobility medium (e.g. buses *versus* trains or metros).

When looking upon public and personal mobility modes from the perspective of 'on the road' conduct, one may argue that riding public transportation vehicles provides room for much more freedom to read, reflect and work, as compared to driving an automobile. This freedom is even more accentuated in trains, which permit moving about the carriages. Thus, following Adey (2006a), we may refer to a relational personal autonomy as far as the mobility experience of individuals is concerned. In any given population we may identify three major sectors regarding the choice between public and private corporeal mobilities: there are those who would always prefer the freedom to read and rest offered by public transportation, there are others who would always prefer to drive their own cars, whereas a third group enjoys the very choice between the two mobility modes, making use of the two modes depending on circumstances.

A significant feature of public mobility is that vehicles are shared by many passengers, so that the lack of personal autonomy, as expressed through vehicle driving, is coupled with a much reduced level of privacy for passengers. This ride-sharing and reduced privacy does not necessarily nullify totally one's personal autonomy, as far as mobility is concerned, but it implies different norms of behavior in a vehicle shared by passengers who are unfamiliar to each other. The loss of privacy may potentially be compensated by the other side of this coin, namely that public mobility provides for potential socializing among passengers. Such potential socializing among passengers is obviously much lower when riding a taxi, whereas when sharing a ride in a private car socializing may depend on passenger/driver relations. Socializing is much more constrained when driving one's own car without any accompanying passengers. However, socializing may emerge also even when driving in solitude, for example when singing along with the radio or CD player, which may be considered as a kind of hybrid between cars and drivers (see Thrift 2004a).

In both public and personal mobilities the simultaneous use of mobility media is possible through the use of mobile phones while driving or riding, and this implies virtual socializing. In public mobility media, such as trains or buses, the

Internet may be used as well. In other words, in physical technologically-based mobility media the more public a medium is the less personal autonomy it permits or facilitates, side-by-side with a reverse tendency as far as potential socializing is concerned (notably when no passengers accompany drivers).

Example 2: Virtual mobility *via* mobile phones *versus* areal mobility

Mobile phones constitute the most ubiquitous personal mobility medium in their penetration: in 2011 there were globally some 87.0 mobile phone lines per 100 inhabitants (ITU 2011)! Furthermore, mobile phones constitute the most instantaneously available personal mobility medium and terminal in the sense of their small size, permitting their being carried constantly by many if not most of their users (see Chapter 7). On the other hand, airplanes constitute the largest mobility vehicles, requiring separate designated parking spaces and terminals at airports and they further constitute a public mobility medium which requires the most extensive preparations for travel, as compared to travel by other corporeal personal and public media. Whether portable or strongly fixed, both media require the use of terminals, and both media require heavy traffic controls, strongly based on IT. Also, both media use the air for their routes though in different ways: mobile phones do not fly and they use the air virtually.

So far, airplanes and mobile phones have been exclusive of each other, since communicating through mobile phones is not permitted while on board, though this may change in the near future. By their very nature mobile phones may be used frequently throughout the day for communications, as well as for other applications installed in them, whereas flying is less frequently performed, and for many mobile phone subscribers or users, notably in developing countries, flying is not an option at all, either because of its high costs or because of people's lifestyles.

Preparations for flying involve heavy use of virtual mobility media, both indirectly and directly. Indirectly, telephone calls may take place and e-mails may be sent to people who may be visited or be met at the flight destination. Such contacts may lead to a decision to fly and they may further bring about additional use of virtual media for direct flight preparations. Another kind of indirect flight preparations involves information sought, notably on the Web, concerning places, events and people to be visited, as well as hotel, and flight schedule information. Direct preparations then take place in the form of flight tickets and hotel bookings bought or reserved online, and boarding passes issued. All these activities are performed at times and spaces other than the flight itself: mainly prior to the flight, at home or in an office. However, they are becoming indispensable components of the flight itself and present a kind of disembedded hybridity between aerial and virtual mobility media. The disembedded nature of this hybridity between aerial and virtual mobilities is somewhat narrowed down at the airport, in which several activities may be performed by passengers via virtual media. These include automated check-in; biometric, machine-operated passport control; machine operated security checks; and the use of boarding passes loaded on mobile phones.

The mobile phone permits ample socialization opportunities, including multi-participant conversations and video calls, and socialization is one of the major drivers for its wide adoption. With the integration of the Internet into 3G (third generation) mobile phones, they permit an initiation of completely new social contacts, through blogs, *Facebook*, *Twitter*, and other networking possibilities, while permitting varied levels of privacy and disclosure of self-identity depending on the choice of the communicating persons (Kellerman 2007). Planes and even more so airport lounges may permit face-to-face socialization, by passengers flying together and potentially also by passengers strange to each other (Kellerman 2008).

Conclusion

We have attempted to show that mobility may be considered as a single entity despite the wide ranging differences among the three major mobility types, terrestrial, virtual, and aerial. This claim was based on the growing conceptualization of mobility as a single social structure. It was further based on several convergences among these three mobility types, as well as on the central role played by IT in all the three of them. Furthermore, we have shown that all the three mobility types, whether looked upon from the perspective of moving people, or when viewed from the perspective of mobility systems, share a basic model of mobility cycles: 'push' and 'pull' effects of mobility motivation, followed by the process chain of IT→origin→route→control→destination. This basic model is built into the very process of movement, and was also relevant in the past to ancient mobility means such as carts and horses or carrier pigeons. However, a conceptual change has typified contemporary technology-based mobilities, beyond just the tremendous increase in the speed of mobilities, namely the ability to transmit information through vehicle-free virtual mobility media. Messengers and postal systems have been replaced by a conversion of information of all types into digital bits transmitted without any vehicle or any other information enveloping component. However, here too the same basic model of mobility cycles applies, consisting of IT, origins, routes, controls, and destinations. Therefore, it will be possible to assess mobility modes and technologies which will be developed in the future in light of this same model, and thus compare them to existing mobility modes and technologies.

In short, mobility has become fully IT-dependent (see Kellerman 2000), but this dependency is a one-way relationship, so that the operation of IT is independent of mobility and is, thus, widely integrated and used in almost all sedentary as well as mobile life segments. However, the extremely wide penetration of IT into mobile and sedentary sectors and activities may imply some, still unknown, future connections of virtual mobility with other spheres of life, beyond the already existing relations with several sectors and activities through, for instance, e-learning, online shopping, home-based business/work, home/online banking, and e-government.

Viewing mobility as a single entity calls for some intriguing future research, both conceptually and empirically. Mobility as a single entity permits analysis of all the movements of a person along a day and/or a week, including both corporeal and virtual mobilities, all of which can be examined along the chain process of motivations, origins and destinations, 'on the road' controls and conducts, and the length of stay in each destination. Such analyses could amount to modified versions of Hägerstrand's classical time geography (Hägerstrand 1970, see also Couclelis 2009). Such time-space examinations of human actions may shed some light on the preferences of individuals, their choice of mobility modes, not only at each origin/destination, but also while on the road. Furthermore, processes and patterns for transitions in modes of behavior may be examined in light of changing mobility modes throughout a day/week. At the level of mobility systems, comparative attention may be given to control processes used for the various mobility technologies and modes, trying to assess whether all of these controls are of the same kind.

Contemporary life points to mobility as a single entity, since moving physically while being engaged in virtual mobility has become routine, and since the very use of mobility technologies has become a basic and even banal activity. Thus, even the common and rather convenient academic division and separation of mobility into categories is now somehow artificial from the perspective of the daily experiences of citizens of globalizing society using simultaneously several mobility media.

The increased availability of diversified media of mobility and their growing centrality in the lives of contemporary individuals have not yet yielded all their potential fruit, in terms of e-learning, e-commerce and various other applications (Kellerman and Paradiso 2007, see Chapter 11). By the same token, the increased varieties of mobility media and their expanding use could have potentially called for better understandings among human beings as individuals, as well as among social groups, notably among nations. That the latter has not yet emerged may point, among other issues, to the importance of national cultures and languages even in a globalizing world, as well as to the crucial significance of the daily physical-spatial context of human life, not yet significantly substituted by other physical and virtual contexts. The immense power of contemporary, IT-driven and integrated, media for daily spatial mobility have still a long way to go to realize their potential for social change.

PART II
Daily Mobility Types

Chapter 6
Terrestrial Daily Mobilities

There is no space we can recognize without some kind of flow – light, sound, people, diseases, commodities, messages, and so on. Without flows space simply would not exist since seeing requires a flow of light, hearing requires a flow of sound, and the lived experience of space requires a flow of human bodies among various locations.

(Adams 2009: 44)

This chapter moves us into the second part of the book. Following the exposure of the numerous roots and triggers, as well as the nature of daily spatial mobilities at large, treating jointly all three mobility types in Part I, this second part of the book will have us explore the very performance of daily spatial mobilities separately for each of the three major mobility modes: terrestrial, virtual and aerial.

This chapter, devoted to terrestrial daily mobilities, will focus on the major modes of terrestrial–corporeal daily mobilities: public transportation; walking; car driving (automobility); and cycling. Since daily mobilities are the focus of the book, most of the attention will be given to public and personal mobilities within urban contexts, and, hence, each of these mobility modes will be examined separately within its proper urban framework. Public mobility consists mostly of electric and petrol buses of various sizes and types, light trains (trolleys), metros, trains, and to some degree also taxis. It may further include bus boats in places where terrestrial buses cannot function (such as in Venice), and ferries which carry private cars and their drivers from one road infrastructure to another. Personal mobilities will include walking in cities, motorcycle riding and cycling and private car driving. The emergence of all of these modes of corporeal mobility will be examined, as well as users' tendencies regarding their use and their social-cultural significances.

The media for public and personal mobilities are referred to in transportation geography as 'collective transportation (public transit)' and 'individual transportation' respectively. Furthermore, it was noted that generally speaking each transportation/mobility mode/medium may fit different distances, with walking fitting the shortest distances and trains and cars the longest (Rodrigue et al. 2006).

Sheller and Urry (2003) discussed the concepts of 'private' and 'public' and referred to several publics and privates, avoiding though a discussion of public and private (personal) mobilities. More generally they claimed that "people move within and between the public and the private, at times being in effect in both simultaneously" (p. 115). As a matter of fact, public transportation is not fully public, in the same sense as city streets are conceived of as 'public sphere' or 'public domain', since entering any public car requires payment through a ticket or otherwise for a kind of short-term 'rental' of moving space over a set distance. Public transportation is, thus, public in the sense that movement on its cars is shared by many riders who buy the same license, namely a ticket. Driving private cars does not require ticketing, but it implies the purchase of the vehicle and the use of an entrance key which plays a similar role to tickets in public transportation. Passengers in private cars need only oral permission by the driver, in lieu of 'ticketing', for entering the private domains of car owners. Both private and public cars require governmental licensing and fee/tax payment in order to permit them to access and use the public infrastructure of roads, whereas pedestrians may use this infrastructure free of charge. The public sphere of road infrastructure is, thus, not fully public for mobility vehicles since its admission and use are not freely available for them even if the fee is paid annually rather than before each particular use of the road system.

Public mobility media

Public terrestrial mobilities, based on public transportation systems, usually share road infrastructures with personal mobilities, which are based on cars and cycles, but they possess exclusivity in railroad systems which cannot be used by personally driven cars. Still, the very construction of both road and railway systems does not yet promise dense and frequent public transportation services, since the latter depend on transit service operators, both governmental and commercial ones, the service frequencies of which are based on governmental policies as well as on public demand for their services. Thus, whereas personal mobilities depend only on governmental road infrastructures and on mobile agents, public mobilities depend also on mobility mediators, namely the service operators and their team of drivers.

Horse-driven public transportation already existed in Europe in the seventeenth century, side-by-side with gradually emerging new public spaces for walking, but "the nineteenth century can [thus] be described as the century of 'public mobilization' through new times, spaces, and sociabilities of public movement" (Urry 2007: 91), side-by-side with the new technology of steam engines, powering locomotives which carried cars with large numbers of passengers. This *public mobilization* by trains was typified by new *connectedness* of people and places, in their provision of fast and comfortable mobility (Urry 2007). The train, as the first technology-based medium of terrestrial (public) mobility, brought about, at the time, the very introduction of several of the basic dimensions of terrestrial mobility, some of which were elaborated in Chapter 1: timetables, and speed as a social value (Virilio 1983, 1992; Knowles 2006); the compression of time-space (Harvey 1989); and new sociabilities at both railway stations and inside train cars. Train cars provided for the first time walkable in-car spaces, thus bringing about a kind of a 'democratic' social environment, though divided into class compartments (Schivelbusch 1986, Urry 2007, see also Kellerman 1989).

Three major technologies emerged in the 19th century for the transformation of terrestrial mobilities into mechanized ones: steam engines in the early 1800s, and electric power and internal combustion engines in the latter quarter of that century. Interestingly enough, all of these innovations were applied to public transportation soon after their introduction, though in different ways. Steam engines, powering locomotives which run over tracks, could not be applied to personal mobilities by their very nature, so that soon after their development, they were used for public mobility service, first in England in 1830, and first aimed at high-class travelers. Electric power and the internal combustion engines were invented in the late 19th century and were soon after applied to public transportation. Electric power was applied to trolleys, trolley busses, trams, and underground metros as of the late 19th century on an experimental basis, followed by full service in the early 20th century in various European and North American countries (see e.g. Smerk 1992). The need for cars to be continuously connected to electricity wires prevented, at the time, the use of electric energy for personal vehicles. Recent technological developments show a reverse trend. Commercially produced private electric cars, powered by batteries, are on experimental sale now and are about to be commercially introduced in 2012 in several small countries (Denmark, Singapore and Israel), but the introduction of electric buses running flexibly through battery power, rather than through connection to street electricity lines, will have to await the development of more powerful and lighter batteries. The third technology for terrestrial mobility, the internal combustion engine, was originally introduced through private cars in the late 19th century, but soon after the first taxis became available, followed by the first buses.

Public transportation is not the most used means of transportation, even for the most common daily and rather pressing travel for commuting. In the US, the flagship of car ownership, only some 5 percent of commuters used any kind of public transportation means in 2008, but this percentage represents a continued slight increase since 1989 when the percentage stood at 4.6 percent (declining by then from 5.1 percent in 1985) (US RITA 2010). In 2008, some 53 percent of the public transit journeys in the US were made on buses (American Public Transportation Association 2010), but as Pucher (2004) commented, the share of the various rail systems is higher when transit passenger miles are counted, rather than rides, given the longer distances normally travelled through them as compared to buses. Low percentages of public transportation are common also in Australia and Canada, but they reach higher percentages in other parts of the world, foremost in Europe. Thus, in the UK in 2006 some 9 percent of the distance travelled nationally was by public transportation as compared to 80 percent by cars, and these percentages remained fairly stable since 1995 (UK Office of National Statistics 2010). Switzerland maintained the highest share of public transportation in Europe in 1995, standing at 20 percent (Pucher 2004). The higher European percentages of public transportation were attributed mainly to more centralized structures of urban areas and higher priorities given by governments to public transportation systems (Pucher 2004).

The contemporary use of public transportation for intra-urban mobility may, first, reflect choice by mobility agents whose 'motility package' includes both personal and public transportation media. As Kaufmann (2002) has shown, preference is not always given to the use of private cars under circumstances of choice, because mobility preferences may depend on commuting distances, on gender as well as on family considerations (see also Canzler 2004, Limmer 2004). However, the very use of public transportation by many others is not a matter of choice, as buses and metros might constitute the only available means for them for mechanized terrestrial mobility, since the purchase and maintenance of private cars might be too expensive for them. The exclusive use of public mobility may be heavily influenced by mobility politics reflecting power geometries (Massey 1993, see also Boltanski and Chiapello 2007, Wolff 1993), leading to the emergence of 'mobility rich' and 'mobility poor' sectors (Cresswell 2006a) (see Chapter 2). Thus, people who depend on public transportation may, as compared to those who have mobility choice, suffer from less access to certain facilities and work opportunities; they may have to spend more time commuting; and they may suffer from a rather restricted ability to move big or heavy objects from place to place. In cases of hazard emergencies, such as Hurricane Katrina in New Orleans in 2005, their ability to escape may turn out as much more restricted, as compared to residents equipped with private cars, ending up with more casualties among the 'mobility poor', who might be racially differentiated from the 'mobility rich' (Cresswell 2006a).

Local mobility planning and politics may worsen conditions for the 'mobility poor' if more resources are directed towards the development of road infrastructure for personal mobilities rather than for public mobility services. Such priorities may encourage urbanites to make more use of more heavily private cars, thus causing public transportation operators to reduce the frequencies of their services, and thus bringing about even lower access for the 'mobility poor' segments of society (Massey 1993, Henderson, 2004, see also Hubbard and Lilley, 2004). Public transportation routes and schedules may frequently reflect general commuting patterns and less the special needs of mothers who have, for example, to travel to daycare centers, to leave their work places earlier than other workers, or to travel shorter distances when commuting (Blumen and Kellerman 1990, Wajcman 1991). Sometimes the order of priority of public transportation schedules may disregard commuting needs, notably when lines are geared foremost to tourists, such as the streetcars of San Francisco and New Orleans.

We discussed in Chapters 3 and 5 the individual and social experiences involved in riding public transportation vehicles. Sharing mobility vehicles with numerous unknown fellow riders may sometimes imply indifference to others, at some other times it may bring about suspicion of the other, but in some cases it may bring about socializing among public transportation passengers. The same wide spectrum of feelings and behavior may apply also to personal space within vehicles, whether buses or train/metro cars. These vehicles might be overly crowded, reducing personal space to zero, or they might be highly spacious if only a few passengers share a specific ride. Public transportation permits moving about the cars and it further permits a wider range of activities while travelling: reading, reflecting and working. Reading for pleasure and work on the train were shown to be the preferred activities during train journeys (Urry 2007), and this is impossible when using any of the media for personal mobility since all require either driving or careful walking. The second preferred activity on train journeys was making phone calls and this mobility activity is possible and much done also in all forms of personal mobility. Mostly, the use of public transportation requires careful planning per line schedules by passengers. Such planning may normally mean longer time spent on the road as compared to car driving, though this may not be the case as far as fast metro lines are concerned, avoiding road traffic jams.

Personal mobility media

It was possible, in the previous section, to jointly discuss the major media of public transportation, namely various forms of buses and trains, as they all share both similar principles of organization and the experience of shared mobility. However, when looking, in this section, at personal mobility, defined as "moving of the self by the self" (Kellerman 2006a: 1), separate discussions are needed for each mobility medium: the natural and non-mechanized walking; the highly mechanized cars with their cabins for drivers and passengers; and the mechanized and non-mechanized media of outdoors cycling (see also Kellerman 2006a).

Walking

"In terms of the history of movement, walking is easily its most significant form, and it is still a component of almost all other modes of movement" (Urry 2007: 63). Albeit, walking is the declining form of mobility in general and of corporeal mobility in particular, probably with the exception of walking as an exercise for physical training. This is evident from data for various countries for the 1990s. It was estimated for the US that merely 10 percent of daily trips constitute either walking or cycling (Dunn 1998: 187). In European countries the percentages were much higher but still declining significantly. Thus, in the UK, the total journeys per person per year increased from 1975/6 to 1994/6 by 14 percent, but walking journeys declined from 35.9 percent to just 29.4 percent of all journeys, and cycling declined drastically from 3.3 percent to just 1.6 percent, respectively (see Stradling *et al.* 2001). Measured in percent of daily travel time, just 16–18 percent was devoted to walking in Sweden in 1990/1, and some additional 3–7 percent to cycling (Vilhelmson 1999). However, there are also exceptions to these trends, typifying bustling cities. It was estimated that "two-thirds of all journeys around downtown and midtown Manhattan are still made on foot" (Solnit 2000: 188).

Obviously, walking as mere locomotion, or as "everyday movement in space" (Seamon 1980: 148), is older than human history, and in this regard it represents pure corporeal mobility leading one from origin to destination. It was simply defined as: "any spatial displacement of the body or bodily part initiated by the person himself" (Seamon 1980: 148). However, Solnit (2000: 14), in her detailed exposition of *Wanderlust*, claimed that "walking as a conscious cultural act rather than a means to an end is only a few centuries old in Europe", with 18th century Rousseau one of the first famous walkers (see Urry 2007 for a detailed history of walking). It is in place, therefore, to distinguish between pedestrians at large, on the one hand, and 'walkers', on the other, the latter constituting a special class of pedestrians. Thus, "Paris is the great city of walkers. And it is the great city of revolution" (Solnit 2000: 218). In his interpretation of walking, notably in New York, De Certeau (1985) preferred the German expression of *Wandermänner* for walkers, "whose bodies follow the cursives and strokes of an urban 'text' they write without reading" (p. 124). Walter Benjamin's solitary *flâneurs*, constituted a special type of Parisian walkers, associated "with leisure, with crowds, with alienation or detachment, with observation, with walking, particularly with strolling in the arcades" (Solnit 2000: 199). Solnit (2000: 27-8) lamented the lack of treatment of walking by postmodernism, which has rather preferred travel for its accent on mobility. Thus, for postmodernists "the body is nothing more than a parcel of transit...it does not move but is moved" (p. 28). As we will see in the next sub-section, contemporary modernity has accentuated autonomous personal corporeal mobility *vis-à-vis* automobiles, as a distinct form of travel, and walking, as a declining form of mobility, remained behind.

Walking may mean much for individuals. It constitutes "the most 'egalitarian' of mobility systems" (Urry 2007: 88). Walking *per se* implies a lack of site (De Certeau 1985), but it requires a destination (Solnit 2000: 249). Repeated walking through well-known streets and paths may reaffirm a sense of dwelling (Urry 2000: 141), whereas any walking provides for freedom and pleasure, "free time, free and alluring space, and unhindered bodies" (Solnit 2000: 173; 250). Beyond these values for all walkers, rural walking involves also a love for nature, whereas urban walking is more like travel, and thus involves danger, exile, discovery, and transformation (Solnit 2000: 188, see also Urry 2007). Furthermore, urban walking is complex and diversified as far as individual motivation is concerned, ranging from soliciting; cruising, promenading, and shopping, to rioting, protesting, skulking, and loitering (Solnit 2000: 174).

Walking requires also some strategic thought involving discipline and regimentation, as for the proper activities, paths and forms of performance, depending on time availability, as well as on social and spatial barriers, notably when walking in cities. These considerations may give rise to some tactical behavior for the seizing of opportunities for safer, faster, or more pleasant walking (De Certeau 1985, Urry 2000: 57). Automobility, which will be discussed in the following subsection, constitutes the individual and societal aspects and significances of the wide purchase and use of automobiles. When compared to automobility, walking is slow and does not provide a sense of power. It is much more limited in its spatial extent, but does not require any dependence on services, and is less regulated than driving. On the other hand, it provides for flexibility, though sometimes less than that provided by cars because of its lower speed. It involves pleasure, though, again, of a different nature than driving. Walking, like driving, provides for personal autonomy and individualism, but in more restricted ways as compared to moving about the city inside a private and comfortable room-like box space of a car (see Chapter 3).

It has been questionable, whether walking on the street, being directly exposed to fellow walkers, encourages or discourages any kind of social contact. Simmel (1969, see also Allen 1999: 63), as well as Solnit (2000: 186) argued for a street of strangers, in which visual impressions and glances are meaningless, but for Jacobs (1961, see also Allen 1999: 63), on the other hand, chance encounters and the very co-existence of pedestrians on the street are meaningful. It seems, thus, that changing mobility modes from public mobility through bus riding or through train travel to walking, or the other way around, involves a rather minor change in terms of human encounters, whereas changing from personal mobility through car driving to personal mobility through walking is more significant, given the difference in the spatial settings and physical exposure to the environment of mobility agents using these two mobility activities.

Automobiles imply status symbols and their adoption and use a democratic right, but walking touches upon more basic democratic values. Thus, walking may be interpreted as an initial step towards citizenship, since it widens the social and spatial microcosm of individuals into a rather wider one. Furthermore, many walking activities constitute integral components of democratic and public life. Such are political and religious processions and marches; military, ethnic or commemorative parades; street parties; demonstrations; and uprisings by people who know the city from their previous walking experience (Solnit 2000: 176, 216–7). At yet another social level, city walking and city streets have seen major changes in more recent history, as far as women's equal, non-sexual, integration in them (Solnit 2000: 234).

Walking contributes significantly to urban space. De Certeau (1985, see also Urry 2000: 53) considered walking as another kind of human speech using space as language. Walking for him is "a space of utterance" (p. 130). Walkers speak and write texts through the space language, as "walking affirms, suspects, guesses, transgresses, respects" (p. 132). Solnit (2000: 213) noted that this language may have turned into a dead one, because of increased automobility. Widely-spread suburbs have meant growing distances among people as well as between people and their daily destinations, thus repelling walking as a preferred and useful personal mobility medium. Sometimes, widely-spread suburbs of single-dwelling houses simply avoid walking by a lack of sidewalks. An opposite trend of shrinking open space may be observed in downtowns, where widely open space has frequently turned into merely minimal space separating among high-rise buildings (Solnit 2000: 175-6, 253-4, 259). Walking in the urban context vitalizes space through the creation of human movements, and it further maintains space as a 'living' entity (De Certeau 1985, see also Urry 2000: 53, Solnit 2000: 176).

Walking was viewed as combining landscapes and walking humans (Ingold 2004, Wylie 2002) and as involving senses, such as sight, hearing, smelling, etc. (Adams 2001, see also Adey 2010). However, within contemporary urban spatial contexts walking has to be related to the more dominant automobility, which may bring about some walking (e.g. from and to cars) while simultaneously reducing or avoiding other walking. Thus, Thrift (2004a: 44) argues "that much walking, both historically and contemporarily, is derived from car travel...in concert with the evolution of automobility", thus implying "other languages which also have something to say". City space or language is now dominated by cars, both moving and parked, so that the visuality of cities is automobilized. Furthermore, city noise and smell are dominated by cars, other than in parks and areas forbidden for car transportation. These noise and smell are completely different in volume and constitution than those of horses until a century ago. From the perspective of pedestrians, or city language speakers and writers, coming into the city by cars, walking through car-dominated landscapes to a destination, and then leaving the city by car, may turn walking itself into a car-dominated and car-mediated activity. The city language may then turn into a rather automobilized language which may also include a necessary but minor component of walking.

Automobility

Traditionally, the US has been considered the country with the highest household penetration level of car ownership, reaching some 89 percent in 2007 (Nielsen 2007). In the past, the US was followed by major European countries in the household penetration levels of car ownership. However, the 2007 data ranked Saudi Arabia as second to the US with a penetration level of 86 percent, followed by the UK (80 percent), Germany (76 percent) and Korea (74 percent). It is, thus, that the wide adoption of automobiles has spread into global regions other than North America and West Europe. As far as the number of cars at large per population it was rather for the Netherlands Antilles to rank first in 2007 with close to 1,200 cars per 1,000 population, followed by Monaco and the US (IRF 2009). Automobility has, thus, become a major mobility experience for a growing number of people all over the developed world.

We mentioned cars and automobility in the previous sub-section in lieu of walking. We will focus now on the major aspects automobility *per se*. "One of late modernity's most recognised and contested objects is the automobile" (Beckmann 2001: 593), "often viewed as the avatar of mobility" (Thrift 1996: 272). As mentioned above, the automobile was invented and introduced during the last quarter of the nineteenth century, and it was first industrially produced in 1908, when Ford's T-model was first marketed in the US (for a bibliography on the history of the automobile see Fischer 1992: 331, n. 55, see also Urry 2007). As we have noted already, in contemporary developed societies the automobile has become the most widely used medium for personal physical mobility.

The very term *automobility* was viewed as consisting of autonomy and mobility (Stradling *et al.* 2001, Featherstone 2004), but for Urry (2004b: 26) it represents more than that:

> The term 'automobility' captures a double sense, both of the humanist self as in the notion of autobiography, and of objects or machines that possess a capacity for movement, as in automatic and automation. This double resonance of 'auto' demonstrates how the 'car-driver' is a hybrid assemblage of specific human activities, machines, roads, buildings, signs and cultures of mobility (Thrift 1996: 282–4).

Freund and Martin (1993) referred to automobility as an ideology, consisting of freedom through power and speed, individualism, pleasure and sexuality. Sheller and Urry (2000), on the other hand, viewed automobility rather as a condition, expressing *auto-ness* in a double sense of human-self and moving machines. Automobility, thus, constitutes "a complex amalgam of interlocking machines, social practices and ways of dwelling, not in a stationary home, but in a mobile semi-privatized and hugely dangerous capsule" (p. 739). Sigmund Freud (1930) viewed the automobile as an 'accessory organ', "an exo-skeleton, a travelling whelk-shell from which, in safety, to contemplate the surrounding monsters; an extension of home, a refuge from the mob, a private cave of autonomous comforts" (Brandon 2002: 4).

The numerous aspects which comprise automobility have received wide attention in various writings (see e.g. Sheller and Urry 2000, Fischer 1992, Freund and Martin 1993, Fischer and Carroll 1988, Gorz 1980, Kern 1983, and Urry 2007). One of the most striking attributes of automobility for individuals is obviously the speed of movement, as compared to any other mode of terrestrial mobility. Related to this speed of movement is flexibility in physical movements, assuming a good and dense road system available to moving agents. As such, automobility implies a higher level of personal autonomy and individualistic mobility when compared to walking and cycling. The very capability for persons of 'mastering' distance and space enhances individuals' power, on the one hand (see Swyngedouw 1993), and their pleasure, on the other. These basic features of car driving, namely power and pleasure were found as most decisive in people's preferences to drive (Stradling *et al.* 2001). On the other hand, driving also involves fear, frustration, pain, and envy (Sheller 2004a, Stradling *et al.* 2001).

Private car driving implies coercion by laws and regulations concerning both the driver and the use of the road system. Sometimes, driving may further require some flexibility in routine time organization, as possible delays may be caused by traffic jams. This aspect has currently been relieved to some degree through the widely spread adoption of mobile phones, permitting drivers' advance reporting of possible delays in their reaching of a destination while being still on the road, as well as more flexible time coordination in general (Larsen *et al.* 2008). Automobile ownership and use may also bring about a reorganization of daily lives, as new facilities become reachable as compared to access through public transportation. As mentioned already, car owners tend to assess public transportation through the lenses of car driving, and preferences for either of them may be differentiated by commuting distances and gender (Kaufmann 2002, Canzler 2004, Limmer 2004). The supposed freedom and increased autonomy associated with automobility are restricted by dependence on maintenance requirements, as well as on petrol supply and prices, as the automobile is the most significant form of natural resource consumption (Urry 2003a: 68).

At the societal level, Urry (2002: 190) identified a "civil society of automobility", assessing automobility as a democratic right, permitting free movement: "mobility by car has become a human right" (Hägerstrand 1992: 36). High adoption rates of private cars may further bring about an emerging culture of automobility, suggesting norms and codes for both road behavior and car looks (see Kellerman 2006a). By their very size, their constant use of the public sphere, and their containing of the human body and kind of expanding it, automobiles have become status symbols, with striking differences among cars in size, shape and color. Cars as moving cabins may make their drivers or riders feel a sense of what Urry (2007: 124) termed 'dwellingness', as if the car is a kind of dwelling space.

Automobile ownership and use may have social ramifications in numerous dimensions of life. Thus, for example, the very ability to move freely by car may involve lower localism, as people may live in one community and socialize in another to which they may go conveniently by car. Fast travel between cities has further promoted *placelessness*, in that differences between cities have been blurred, and drivers and passengers alike may disregard landscapes and places while on the road between origins and destinations (Relph 1976). Yet another dimension, the assessment of the social significance of automobile ownership went as far as claiming that "automobiles enhance mobility, and mobility enhances knowledge" (Lomasky 1997: 16).

Spatially, automobility implies that a major chunk of urban areas is devoted to cars, mainly for roads, parking lots, and maintenance facilities. In some automobile-dependent cities such as Los Angeles, car related functions constitute the largest urban land-use, with almost one-half of the urban area devoted to car uses (Urry 1999). Automobiles have further facilitated the suburbanization and sprawl of population, services, and industrial production, side by side with growing residential segregation, facilitated by car ownership. Some viewed such processes as an "irresistible replacement of slow spaces by faster ones" (Hubbard and Lilley 2004: 277). The extensive use of growing numbers of cars has been accompanied with extensive negative environmental effects, notably pollution and noise.

Looking towards future developments, Urry (2004b) identified six transformations which, in his opinion, may lead corporeal mobility into a new, so-called, *post-car system*. First, new fuels, notably electricity, have been introduced, and second, new materials for construction, e.g. aluminum, have been incorporated into car production. A third transformation constitutes smart-card technologies which permit higher levels of hybridization. Fourth, there are various initiatives, such as car sharing, which may bring about de-privatization of car use. As a fifth transformation one may view the more comprehensive transportation planning which integrates car-based roads with public transportation, as well as with walking, and land-use planning. Finally, the sixth transformation refers to the integration of transportation with information technology, both in the sense of information technologies as providers for a better functioning of automobiles, and in the sense of communications technologies offering some substitution for corporeal mobility. Some of these changes are either still experimental or they are limited in use, such as new fuels, whereas some other changes are still awaited, such as the substitution of transportation by communications. A possible future full materialization of these transformations will not necessarily present a post-car era, because cars may still constitute the most important medium for physical mobility even when new mobility technologies will become available. A new mobility era might possibly amount to the introduction of rather integrated and smart mobility media, which may bring about a sophisticated and more careful use of environmentally friendly cars, side by side with a more efficient division of movements among various physical and virtual mobility media.

Cycling

Cycling was initially introduced as a mobility alternative to trains, providing autonomous, seemingly speedy, and timetable-free mobility, notably for women (Sachs 1992, see also Urry 2007). Following the introduction and adoption of cars, cycling has received additional values beyond being merely another technology for movement from origins to destinations. The open connection between body and landscape while cycling provided for what Spinney (2006) termed 'hybrid rhythms', and furthermore, "a focus on the perceptions and movements of the cyclist can excavate contextualised everyday meanings by illustrating the dialectical relationship between place, practice and representation" (pp. 714–5).

Amsterdam might well be the only big city worldwide which offers three distinct traffic systems for three distinct vehicles for corporeal personal mobility: roads for automobiles; special lanes for bicycles; and canals for private boats. It seems that the canals are not so much in use for personal mobility on weekdays because of the lower flexibility of boats as compared to bicycles and cars. Cycling, however, is most widely used. Nationwide in the Netherlands cycling is second to car driving among commuters (Statistics Netherlands 2004), and it is the most widely used transportation mode within inner Amsterdam (Bertolini and le Clercq 2003). In a country of 16 million people and some 7 million households there were over 13 million bicycle owners (Statistics Netherlands 2004). Furthermore, there might probably be many more bicycles in the Netherlands, up to some 25 million, because many people own simple bicycles for weekdays and more elegant and sophisticated ones for weekends. An extreme domination of a particular mobility technology is pertinent in other countries, as well. Thus, the US used to have more cars than people, and in many countries there are currently more mobile phones subscriptions than people.

Rome is similar to Amsterdam in that the local habit calls for the use of a vehicle other than cars for personal mobility, namely the *motorino*, or the light motorcycle. Like bicycles, *motorini* constitute vehicles which are easy to maneuver and park under heavy urban traffic conditions. Contrary to Amsterdam, though, there are normally no separate lanes provided in Rome for *motorini*, and contrary to bicycles, *motorini* make a lot of noise and contribute to air pollution. The popularity and domination of *motorini* in the corporeal mobility scene of Rome seems obvious when looking at the Roman urban landscape, but data seem less impressive compared to the cycling popularity Amsterdam. By the end of 2004 there were some 1,847,258 million automobiles registered in the city of Rome as compared to some 287,499 motorcycles, or 13.5 percent of the total for the two personal mobility media. As expected, for the whole Province of Rome the share of motorcycles was lower standing at 12.4 percent (for 2,583,009 automobiles and 366,850 motorcycles) (ACI 2005). *Motorini* might thus be a preferred medium of mobility for commuting, but it is used alongside cars which are preferred for other movements.

Conclusion

A wide array of terrestrial mobility media are available for contemporary mobility agents, notably in developed economies, whether public (trolleys; light trains; metros; trains; buses of various sizes) or personal (walking; motorized and non-motorized cycling, and cars of various sizes). As we could note, individual mobility operations through any of the three media for personal mobility, whether old (walking) or newer (cycling and cars), has involved the development of 'mobility cultures' in terms of preferred uses, on the road conduct, and cultural symbols and meanings attached to them. It is difficult to point to equivalent 'mobility cultures' for public transportation media other than long distance, non-daily train lines (e.g. the 'Orient Express' or the Trans-Siberian line). Self-mobility, with or without a mobility vehicle, provides the necessary arena for some personal taste for the very action of mobility, something which is more difficult to experience or achieve when sharing a paid-for movement with many others, all being moved by a professional driver.

It turns out that in most cases whenever it is possible economically for mobility agents to purchase and maintain cars, this is the preferred medium of mobility, even if public transportation may be used for commuting. Cycling is a preferred mode mainly in specific cities or cultures. The car as an individual, personally dominated and sheltered, mobility medium, is highly attractive in contemporary developed society. The tremendous development of personal virtual mobility media, which we will review in the next chapter, is yet another dimension of daily spatial mobilities of a personal nature, as part and parcel of a society dominated by individualism.

The huge newly industrialized economies of China and India have developed cheap cars in order to permit their wide purchase by the newly-formed middle classes. Thus, Urry (2007) did not predict a shining future for public transportation systems. Rather, car systems are now undergoing experimental transitions notably as far as their fueling is concerned (mainly through hybrid, gas, and electricity), in order to reduce environmental, economic and political effects of the current petroleum-based fueling systems, while still keeping the car as the spatially and socially dominant medium for terrestrial mobility.

Chapter 7
Virtual Daily Mobilities

It is increasingly difficult to find spheres of daily life in cities in the Western world that are not mediated in one way or another through such information and communication technologies (ICTs) as the Internet or mobile phones.

(Schwanen *et al*. 2008: 519)

This chapter will attempt to highlight virtual daily mobilities from several perspectives: the qualities and significances of the various communications media (telephone; Internet; and mobile technologies); the nature of cyberspace and its cognition; and the contemporary instantaneity of broadband services. As compared to our discussion of terrestrial mobilities in the previous chapter, the three major technologies/devices which we will introduce in this chapter for virtual mobility are all for personal mobility, since the older public mobility technologies of telegrams and letters have lost their significance and popular use with the development of personal, instantaneous and universally available communications technologies. It is, therefore, important to highlight cyberspace and broadband services, in addition to the discussion of available technologies, in order to understand the bases of operations of contemporary communications technologies, as well as their experiential significances for users.

Back in 2007 Arminen claimed that "the current vision seems to be that mobile communication converges with the increase in local wireless networks. This will lead to a development not foreseen some years ago…The convergence will merge the Internet and the mobile telephone as we know them now" (p. 435). Recent developments and innovations in ICTs and their wide and growing adoption have brought about this merge and have changed the status of cyberspace *vis-à-vis* physical space, as well as its status for users of information and communications devices. The most important change regarding cyberspace has been its permanent and instant availability to users. This 'natural' access and use of cyberspace has brought about its becoming integrated and converged with physical space, at least from the perspective of users enjoying mobile broadband terminals and transmission systems.

The chapter will begin with an exposition of the rather veteran telephone, followed by discussions of the Internet and mobile communications technologies and appliances. We will then focus our attention to cyberspace, on which the Internet is based, suggesting the classification of cyberspace into two basic classes: information and communications spaces. This discussion of the basic classes of cyberspace will be followed by an examination of the more advanced virtual daily mobilities: mobile broadband media and their significances. Finally, we will suggest several implications of instantly accessed cyberspace by mobile broadband users, access which is available without location and time restrictions.

Telephone

Telephones constitute the most basic and oldest appliance and medium for personal virtual mobility (Kellerman 2006a). However, the introduction of new communications technologies, notably mobile phones, has brought about declining penetration rates for telephones in developed countries and low penetration rates in developing ones (Table 7.1). Like the automobile, the telephone was first introduced in the last quarter of the nineteenth century (1876), and its history has been documented elsewhere (see e.g. Fischer 1992). In the early twentieth century the facsimile (fax) was introduced in the US as a most cumbersome and space-consuming complementary technology for textual and graphical transmissions over the telephone system, serving at the time mainly newspapers, meteorologists and security forces. Its digitization and miniaturization in Japan since the 1970s has made it a standard office and home device.

In the following paragraphs it is intended to review briefly various sociospatial spheres and aspects of telephone use (see Sheller and Urry 2000, Fischer 1992, Freund and Martin 1993, Kern 1983, Kellerman 1993, 2006a). Whereas the initial mass-production of automobiles was geared to the household market, the early commercial introduction of the telephone was meant for the business one. Moreover, once the telephone began its penetration into households, in the US of the early twentieth century, no significantly distinguished telephone models for home use were introduced. The telephone has always been a small appliance which normally has had to be attached to the body when used, rather than being a large-scale capsule enveloping the body as cars have been.

The telephone implies speed, as it provides speedy and instant two-way communication, qualities which were unattainable for communications through the previously available media of postal services and telegraph. The telephone is much speedier than the automobile, as it does not involve time and space frictions for call transmissions. These features of the telephone took several years of development at the time, as early telephone service required some time-consuming assistance by telephone exchange operators. Like automobiles, telephones provide flexibility in movements, when assuming that most households and businesses are connected to the system. Also like automobiles, telephones provide subscribers with personal autonomy and individualism (see Fischer 1992), at times most significantly for women and children who were partially deprived of these qualities prior to its introduction. Telephones further involve the power of information sharing, as well as the pleasure associated with social contacts. Given the lack of time friction or traffic jams in their normal operation, the telephone may assist routine time organization and, hence, it involved a reorganization of daily lives when services became obtainable over the telephone.

Table 7.1 Penetration rates of communications technologies by world regions, 2009 (in population percent)

Medium/ region	Africa	Asia & Pacific	Arab States	The Americas	CIS	Europe	World
Fixed–line telephone	1.6	14.0	9.4	28.1	26.6	40.1	17.1
Mobile phone	37.5	56.0	72.1	90.4	118.9	127.8	76.1
Internet	8.8	18.4	19.3	35.7	48.3	62.9	30.0
Fixed broadband	0.1	1.7	4.6	6.5	14.3	22.4	8.0
Mobile broadband	2.2	5.4	5.7	16.1	19.1	33.0	13.4

Data source: ITU 2010b.

The major differences for individuals between automobiles and telephones are the lack of laws and regulations for 'driving' the telephone or 'passing through' communications lines, as well as the lower dependence on maintenance and global supply industries (such as petroleum) in telephone usage. Telephone use involves also several features which are irrelevant for physical mobility. Thus, the telephone provides for co-presence, or connected presence (Tillema *et al.* 2010), in at least two places when a conversation is made, and it amounts to "disembodied sounds – of speech displaced in space and time from its origins" (Mitchell 1995: 36). The telephone is, thus, a time intruder when one is being called by another party.

Early governmental initiatives, such as the 1934 American Telecommunications Act, attempted to assure universal availability of household connections to the telephone system and assumed, therefore, that personal virtual mobility constitutes a basic right, even if the provision of the service is not performed directly by government but is rather channeled through private companies.

Informal social relations, as expressed through telephone conversations, may present some special nuances for audio interactions. Thus, telephone contacts have created telephone cultures as to the way telephone conversations should begin and end, as well as for their structuring, time of calling, etc., and these nuances and cultures might differ from country to country. Callers are expected to abide by such domestic 'norms'. Similarly to the driving of private cars, telephone use for social contacts may facilitate increased contacts with other places, and these contacts may bring about some declines in localism coupled with increased placelessness. The wired telephone may be considered a significant social device, but despite its social importance it has not become a status symbol like automobiles, given its being a small appliance located inside homes.

From a spatial perspective, and contrary to automobiles, contemporary telephone infrastructures constitute a minor and declining land-use, as most cables are buried, and digital telephone exchanges becoming smaller and smaller. Until the introduction of digital telephone services in the 1960s, analog telephone exchanges required much space, so that telephone exchanges were present in most urban landscapes. Also contrary to automobiles, telephones have not been considered an environmental pollutant. However, like automobiles, they have facilitated the suburbanization of population, services, and production (see Kellerman 1984 for literature). As such, telephones do not necessarily contribute to segregation, but they may facilitate personal physical isolation while still permitting the maintenance of virtual communications with and by solitary individuals.

Internet

The Internet was originally tried out in the US in the 1960s as an electronic mail network, developed in order to serve as an alternative security network for the telephone and telex systems in case of a nuclear attack (Kellerman 2006a). Its current universal availability has been considered as the best example for the adoption of a technology for purposes completely different than those envisaged by its developers (Urry 2003a: 63). Since the mid-1990s the Internet has been stabilized as including a universal and global e-mail system, namely a communications system, as well as the World Wide Web, which constitutes the largest library and information storage entity worldwide, constituting an information system. The integration of these two Internet components of communications and information into one system permits instant and worldwide availability of both personal and public information (for Internet development see Kellerman 2002).

The Internet was considered to constitute "a metaphor for the social life as fluid" (Urry 2000: 40). It is in many respects similar to its predecessors in mobility facilitation, the telephone and the automobile, in its being a free and uncontrolled medium for personal mobility. Like telephones, there are no laws and regulations for 'driving' the Internet or for 'passing through' communications lines. Also like the telephone, the Internet provides for co-presence or connected presence in at least two places when a real time interaction is made. On the other hand, one may also point to some differences between the Internet and the telephone. If not used for incoming telephone calls (through VoIP), the Internet cannot be considered a time-intruder. In other words, the Internet facilitates its operation by users at any time but it does not force temporal intrusion or intervention into the time of communicated partners as the telephone does through the sounds which signal incoming calls. The Internet is similar, though, to telephones as far as spatial expansion is concerned. Thus, e-mailing permits long distance and international messaging and calling at no charge per call or by time, and the Web expands lived-spaces of its users beyond their real locations, since cyberspace represents additional real or imagined spaces, as we will see in the next section.

Socially, the Internet has expanded the idea of personal virtual mobility as a democratic right by its very provision of instant written communications, as well as through its provision of access to information. It has practically extended personal virtual expression to unprecedented levels, through both the Web, notably through Web 2.0, and obviously through e-mailing. One basic dimension of these extended mobility and speech expressions are the emerging virtual communities and networks. The Internet further permits co-presence in several places simultaneously (Urry 2000: 71). Thus, the Internet involves exposure of its users to virtual spaces as well as to geographically more dispersed social ties, and these may potentially be associated with increased placelessness (see e.g. Dodge and Kitchin 2001, Wellman 2001b). As such, the Internet is closer to automobiles than to telephones, as it permits both audio and visual exposures to virtual spaces.

Virtual and real services provided through the Internet may potentially contribute to an increased and spatially more extensive dispersion of dwellings. Visions calling for future totally dispersed spaces of residence have accompanied earlier phases of telecommunications developments but have not been materialized (Kellerman 1984). So far, service provision through the Internet has not yet shown evidence for the emergence of spatial community dissolution.

The maintenance of the Internet and its spatial effects are completely different than those of automobiles. The use of the Internet involves some dependence on experts in telecommunications and computers, more than for the use of telephones, but still at much lower levels than the periodic and unpredicted mechanical maintenance of automobiles. Contrary to automobile maintenance, computer problems can frequently be remotely fixed or taken care of by users themselves with expert guidance over the telephone. The Internet is transmitted through telephone lines so that it does not add any specific land-uses other than computer stores, ISP offices and facilities, and Internet hotels for website storage servers, etc. Like the telephone, and contrary to automobiles, the Internet adds less to environmental pollutants, mainly through the side effects of electricity production for the Internet hotels.

Mobile technologies

From the perspective of individual users, the most dramatic trend in ICT innovations in recent years has been the Internet becoming fully mobile and universally available, thus permitting access to both e-mail and the Web at any time and any place, through at least two technologies (Wi-Fi and cellular modems). This universal and instant availability of the Internet implies a practical integration of cyberspace with physical space, again from the perspective of individual users, as we will discuss in a later section (Kellerman 2010b).

We may identify four phases in the emergence of mobile information technologies from users' perspectives, beginning with the introduction of basic communications devices, the laptop and the mobile phone, moving through innovations of information and transmission systems, and ending up more recently with the introduction of advanced communications devices making use of the previously innovated transmission systems (Table 7.2). First was the invention of the laptop PC, which was introduced commercially back in 1975, serving at the time without wireless communications. Its current penetration rate has lost some of its significance since smartphones and Tablets can do much of the work of laptops. However, when comparing the popularity of laptops to desktops, it would suffice here to note that in 2008 more laptops were sold in the US than desktops.

Table 7.2 Mass introduction and penetration of mobile innovations

Innovation	First year of mass production	Global penetration rate in % (end of 2009)
I. Basic devices		
Laptop PC	1975	N/A
Mobile phone	1983	67.0
II. Information systems		
Internet	1994	25.9
SMS	2000	N/A
III. Transmission systems		
Wi-Fi	1991	N/A
3G	2001	9.5[a]
IV. Advanced devices		
Smartphone	1993	N/A
Netbook	2007	N/A
Tablet	2010	N/A

Note: [a] Mobile broadband subscription.

Sources: Table: Kellerman 2010b, Data: Kellerman, 2006: Table 4.1, ITU, 2010a.

The mobile phone is an even older innovation than laptops. Mobile telephone technology was originally introduced in 1906 by Lee de Forest who claimed that "it will be possible for businessmen, even while automobiling, to be kept in constant touch" (Agar 2003: 167). The first limited mobile services were introduced in the UK in 1940 and in the US in 1947, followed by commercial introduction in 1979–1983 through the allocation of wave lengths for its operation. The mobile phone has turned out to be the most widely diffused communications device, with a global penetration rate of 87 percent in 2011, and 90 percent of the world population covered by a mobile phone signal in 2009. This high percentage represents actually a much wider availability of mobile phones, given that in developing countries people rent their mobile phones for single calls or SMSs to others as a commercial service. As of 2002 the percentage population worldwide owning a mobile line has been higher than that having a fixed line (ITU 2010a, Table 7.1; ITU 2011).

Penetration levels of mobile phones were found to be related to personal income, notably in Oceania and Asia (Comer and Wikle 2008), whereas the pace of penetration has been higher in small and/or densely populated countries, permitting easier setting up of wireless infrastructure (Castells *et al.* 2007). Mobile telephony has shown evidence for a positive impact on economic growth notably in developing countries (Kauffman and Techatassanasoontorn 2009) with special significance in Africa, since this technology is frequently the only available mobility technology, representing a *leapfrogging process*, in which a new technology is adopted while skipping the adoption of older ones, in this case mainly the telegraph and the fixed-line telephone (Comer and Wikle 2008). Castells *et al.* (2007) noted, in their internationally comparative study, that adolescents and young adults have led in the utilization of mobile phones for SMS communications, since it has been easier for them to adopt this technology and since its use is cheaper than voice calls. They further noted that with growing rates of adoption of mobile phones within given national populations, gender differences in adoption rates decline, with women tending to make more social uses of the technology than men.

Two information systems have become relevant for the two mobile communications machines, the laptop and the mobile phone: the Internet and SMS. The first, the Internet, as mentioned already, was commercially and fully introduced only in 1994, and it was originally meant to be used with desktops, and later also with portable computers connected through fixed wired servers. Only once wireless transmission systems have become available the Internet became mobile, for both its e-mail and Web components. About one quarter of the global population made use of the Internet by the end of 2009 (ITU 2010a). The Internet requires relatively expensive terminals (PCs, laptops, or smartphones), and it further requires literacy for most of its uses, with the exception of visual and spoken information, such as video clips and phone calls. The SMS, on the other hand, introduced in 2000, was originally invented for text and later also for video messages transmitted through mobile phones, and this is still its major application, though SMSs may currently be sent also through computers as well as through some fixed line telephones. SMS service has constituted an integral part of mobile telephony from its outset, and thus did not require any additional dedicated transmission system.

The integration of mobile devices and the Internet into an extensive and comprehensive information system came into being with the introduction of two transmission systems: Wi-Fi (Wireless Fidelity), meant originally for laptops and now installed also in smartphones, and 3G (Third Generation), implying a third generation of cellular technology permitting broadband communications. Though originally introduced in the early 1990s, Wi-Fi was widely adopted only as of the second half of the 1990s, since it required the development of tiny communications components installed first in laptops and later on also in smartphones, as well as numerous fixed antennas, each covering a limited spatial range of reception. Wi-Fi routers are used both within and outside homes, so its gross diffusion rate may be misleading. For late April 2010 there were reported some 295,589 free and pay-for Wi-Fi hotspots worldwide, almost a quarter of which were located in the US (JiWire 2010). The minimal transmission speed to be considered broadband transmission, or high-speed transmission of data, whether through cables or through Wi-Fi, has increased over the years from 64Kbps (kilobits per second) to the current widely used FCC (Federal Communications Commission) one of 768Kbps (ITU 2003, FCC 2009).

The second mobile transmission system, 3G, with its broadband features, made it possible to use the Internet through the mobile phone transmission system. This technology, first introduced in the early 2000s, was available for Internet use by the end of 2009 to some 9.5 percent of the world population through mobile broadband subscriptions (ITU 2010a). Interestingly enough for the study of the growing mobility of information, the diffusion rate of mobile broadband was higher than that of fixed broadband: by the end of 2009 only some 7.1 percent of the world population enjoyed fixed broadband subscription. Mobile broadband subscription surpassed that of fixed by 2008, just seven years following its introduction! However, the share of mobile subscribers out of the percentage world population having mobile phone subscriptions by the end of 2009 (67 percent) was only 14.2 percent, whereas the share of fixed broadband subscribers out of world total Internet users at the time (25.9 percent) was double than this, standing at 27.4 percent (ITU 2010a). Thus, mobile broadband services, are as yet available to just a small portion of the global population, though the percentage of mobile broadband subscribers is growing fast. Moreover, subscribers to 3G services are restricted to the geographical coverage of the service which is not yet as universal as compared to the geographical spread of basic cellular communications (see e.g. for the US, Verizon 2010).

The fast growth rate of mobile broadband subscription may soon make it a dominant technology, but the adoption rate may reach a limit in the developing world, even if we assume potential future drops in the prices of smartphones and broadband subscription, since most of the uses of the Internet require literacy. There are still wide digital gaps in the adoption of mobile broadband connection. Asia, the Pacific and Europe lead globally, but in Asia, for example, Japan and Korea accounted for 70 percent of mobile broadband subscribers in the continent by the end of 2009, whereas the US accounted for 82.6 percent of the subscribers in the Americas (Table 7.1) (ITU 2010a).

The availability of mobile wireless transmission technologies has brought about, at the recent and fourth phase, the introduction of matching advanced mobile communications devices. The smartphone, originally introduced in 1993 and diffusing widely as of the 2000s, has actually developed into an advanced laptop, but including additional features which used to be available only through dedicated devices, such as GPS for road navigation, and high-quality cameras.

The most recent introduction of Netbook and Tablet laptops amounts to a reaction of the laptop market to the smartphone by introducing smaller and lighter laptops, thus closer in size and weight to smartphones.

The adoption rates of mobile technologies are of special importance for the understanding of daily spatial mobilities because they permit full virtual mobility. Thus, special attention has to be given to the ongoing wide global digital divide which accompanies the introduction of new communication technologies and devices and their penetration (Table 7.1). In all the technologies mentioned so far, Africa is the least developed continent and Europe is the most advanced. There are though differences in the gaps between Europe and Africa by technology. For mobile telephony, the most widely adopted technology in Africa, the European adoption rate is 3.4 times higher than the African one, as compared to the Internet for which the European adoption rate is 7.4 times higher than the African one, 224 (!) times higher for fixed broadband, and 15 times higher for mobile broadband. These differences attest to the lack of terrestrial cable infrastructures in the developing world. In all world regions, mobile broadband penetration rates still call for wider adoption, even in leading Europe with 33 percent of its population having already access to it. However, in all world regions the penetration rate of mobile broadband is higher than that of fixed broadband (ITU 2010b)! There are, of course, wide gaps also within global regions, notably in the Americas and Asia, as mentioned already. Digital divides may be substantial also within countries, whether among regions or among population sectors divided by gender or socioeconomic factors (see Castells *et al*. 2007, Graham and Marvin 1996).

Several of the mobile broadband applications have been specifically developed for users on the move. One such development is extended GPS services, permitting navigation on a global scale. Another such application is a new generation of LBS (Location Based Services), providing commercial local information (e.g. on restaurants) to people's mobile phones by their locations. Furthermore, the availability of mobile broadband for connection to the Internet implies more extensive and more instant uses of Internet applications developed originally for fixed broadband communications. Striking such uses are entertainment, permitting radio and cellvision reception, and the view of streaming pictures in form of video clips and full movies, side-by-side with e-work (synonymously called telecommuting or telework), e-banking, e-government, e-health, e-learning (synonymously called distance learning), and B2C (business to customers) c-commerce (synonymously termed online shopping) (see Chapter 11). SMS has turned from a mere interpersonal medium into a business-to-clients medium, as well. The British KAPOW! Survey (2005) was able to rank the top ten business types using SMS: recruitment agencies; entertainment information services; clubs and bars; Internet service providers and hosting companies; couriers; schools, colleges and universities; hair salons, dentists and surgeries; mechanics and body shops; charities; and insurance companies.

All these applications imply not just a widening consumption or use of the Internet by individuals but also the widening and deepening of the production side of the information society, consisting of production and maintenance of additional or enhanced websites, and more indirectly, fostering the consumption of more products and services by individual users of the Internet. The use of mobile broadband applications may widen the multi-tasking pattern which has typified the use of mobile phones while on the road (see Schwanen *et al.* 2008), as well as the coordination of face-to-face meetings (Larsen *et al.* 2008).

The use of some of mobile broadband applications, at home and elsewhere, has tended to develop relatively slowly at the first phase of broadband diffusion, since some of these applications require changes in the supply side of the system, such as legal changes for online shopping, whereas others call for adjustments of demand, such as changes in learning habits needed for e-learning (see Chapter 11). The paces of adoption were recorded for the pioneering and leading Japan and Korea in the mid-2000s, and later on for France which has led Europe in this regard (Kellerman 2006b). By the early 2000s, Japanese customers adopted Internet browsing and e-commerce over mobile phones, at a time when such applications were at their infancy elsewhere, notably in the US. Aoyama (2003) attributed this pattern to a variety of specific social conditions, ranging from market positioning of m-commerce (mobile commerce), through the importance of portability and urban spatial structure, to socially embedded user friendliness attributed to m-commerce (see also Zook *et al.* 2004).

The need for time in order to overcome habits and cultural traditions can be accentuated also by looking at the Internet experience in the country which adopted e-commerce and e-learning first, and which still dominates the scene of e-commerce, the US. The initial advantage of the US, as the country in which the Internet was invented and first widely adopted, was that the ground was prepared for wide advanced uses, even when the US was still lagging behind Korea at the time in fixed broadband penetration. Thus, in 2004 some 50.9 percent of total global e-commerce took place in North America (Hwang *et al.* 2006), and already back in 2000–2001 some 56 percent of the degree-granting institutions in the US offered distance courses, but not necessarily distance full-degree studies. These offerings ranged from merely 16 percent of the private two-year colleges, to 90 percent of the public two-year colleges, and from 40 percent of the private four-year institutions to 89 percent of the public four-year institutions (US Bureau of the Census 2006a). In the currently evolving mobile broadband age, it was reported for 2009 that 21 percent of young US customers used mobile phones for banking and that 25 percent of mobile phone users in general shopped online (Scherr Technology 2009) (see Chapter 11).

Unfortunately, no data on transmission volumes via fixed broadband by Internet general classes of data form (i.e. streaming, downloading, browsing, etc.) could be found in order to compare them directly with equivalent transmissions over mobile broadband. However, the tendency for a fast growing transmission of streaming information has occurred in both fixed, mainly residential, broadband Internet, as well as in mobile one. The large share and growth of this application through mobile broadband in the Pacific, notably in Japan and Korea, has continued the prior trend of video entertainment dominance in these countries through fixed broadband. In the US too it was found that 61 percent of broadband Internet users in 2009 watched online video (Scherr Technology 2009).

It is important to bear in mind that the transmission of streaming information as compared to other applications implies the transmission of a heavy volume of data. Thus, one cannot extrapolate from the type of transmitted information on the time spent by users for each type of application. However, the introduction of Web 2.0 and its social networking, involved not only frequent transmissions of streaming information, but above all a constant sharing of daily experiences and thoughts by millions of people. Thus, on average some 900,000 blog posts were created daily in 2009, some 200 million people or 13 percent of Internet users worldwide used *Facebook* actively in 2009, and one half of them did so at least once a day (Scherr Technology 2009). Broadband subscribers worldwide generated in 2009 on the average some 11.4GBs of Internet traffic per month, which was equivalent to some 3,000 text e-mail messages per day. Korea still led the average use of broadband (24.5GBs per month) followed by France (14.3GBs) and the US (14.2) (Gross 2009).

Combining physical and virtual mobilities by using wireless communications in automobiles or in destinations reached by automobiles, implies complementarity between corporeal and virtual mobility (see Chapter 1). It further implies a reconnection between the movements of people and information, following a long separation between the two, starting with the introduction of the telegraph which enabled the transmission of messages without messengers (see Sheller and Urry 2000: 752, Cooper 2001). From yet another angle, automobiles provide a special moving space of privacy for calls made over mobile phones (Kopomaa 2000: 15). Wireless communications is in many respects similar to its predecessors in mobility facilitation. However, when compared to the telephone it obviously facilitates flexibility in both physical and virtual movements whereas the telephone permits only virtual flexibility. Wireless communications further simultaneously intrudes users' time and space (location), as compared to only possible time intrusion by the telephone. The use of mobile phones, thus, nullifies possible isolation. The use of either mobile phones or wireless Internet connection implies a blurring between the private and the public, as well as between indoors and outdoors. Whereas telephones and computers were traditionally considered devices to be used indoors and involving some privacy of communications, wirelessness implies less privacy and a change of social boundaries regarding the acceptance of communications activity in the public sphere.

Wellman (2001b) views wireless communications as expressing a new phase in social communications and networkings. He termed non-technological communications of people walking for visits of each other as *door-to-door* communications, typifying social relations within traditional, physical-place bounded communities. The automobile and the telephone have permitted the development of a second phase of social relations and networking, *place-to-place* ones, with some flexibility in the location of these places and replacing some of the local door-to-door relations. The Internet has enhanced place-to-place networks through its provision of continuous communications. Placeless wireless communications have implied the emergence of a third phase that of *person-to-person* communications, detached from household location and its communications infrastructure.

Wireless communications devices may constitute status symbols as they present users' "emphasis on coping and continuous movement" (Kopomaa 2000: 14), and in this respect mobile telephones are different from fixed ones. Though being smaller in their size than fixed telephones, mobile ones are being carried and used in public, and the current large variety of features of mobile telephones of various generations, as well as their design and colors, may permit viewing some of them as status symbols.

Townsend (2001) and Zook *et al.* (2004) noted that mobile phones may permit faster, more efficient and flexible use of time and space by their users, which may fit the more flexible social nature of second-modernity cities. They further noted the aggregately more efficient management of face-to-face contacts in CBDs, as well as a more efficient use of highways, when mobile phones are widely adopted, as this communications medium permits immediate contact when, for example, some scheduling requires change because of any unforeseen traffic congestion. Haddon (2004: 96) noted that mobile phones permit more spontaneity in time use. For individuals, wherever located, mobile phones may further imply personal globalization, as overseas destinations may be reached instantaneously from any location, albeit frequently at high costs for vocal calls (and low ones for SMS).

The fast diffusion and adoption of home and mobile communications technologies for personal mobility may imply a decline in public services for virtual mobility meant to serve those who are not equipped with the Internet and mobile phones. Hence, low income people become partially or fully immobile as far as virtual mobility is concerned. This is similar to the processes described in the previous chapter regarding the decline in public transportation when more private cars are purchased. Thus, though the postal service still functions and handles paper mail, telegraph services almost disappeared and telex services are not available anymore in most developed countries. By the same token, the number of public phones in cities declines with the decreasing demand for their use, and Internet cafés tend to be located in migrant neighborhoods.

Cyberspace

The introduction of the Internet in the mid-1990s has brought about a focus on *cyberspace*, a term which was originally proposed by Gibson (1984) as a science-fiction notion, and applied later to computer-mediated communications and to virtual reality technologies (Kitchin 1998: 2). Since the early 1990s, cyberspace has been variously defined as:

1. *Artificial reality*: "Cyberspace is a globally networked, computer-sustained, computer-accessed, and computer-generated, multidimensional, artificial, or 'virtual', reality" (Benedikt 1991: 122, see also Kitchin 1998: 2).
2. *Interactivity space*: "interactivity between remote computers defines cyberspace...cyberspace is not necessarily imagined space – it is real enough in that it is the space set up by those who use remote computers to communicate" (Batty 1997: 343-4).
3. *Conceptual space*: "the *conceptual space* within ICTs (information and communication technologies), rather than the technology itself" (Dodge and Kitchin 2001: 1).

These three definitions may be viewed as complementary rather than contradictory to each other. Thus, cyberspace constitutes simultaneously a virtual, interactive and conceptual entity. The common thread among these three definitions of cyberspace is that cyberspace has been viewed as a category of space or reality. Cyberspace may further represent real space through maps, pictures and graphs, which may then be used for an understanding of real space and for navigating in it (Zook and Graham 2007a). Larsen *et al*. (2006) proposed a differentiation among *imaginative travel* through images and memories, *virtual travel* through the Internet and *communicative travel* via letters, phone calls and e-mails. This differentiation lost much of its practical meaning due to the growing convergence among mobile phones, the Internet and TV.

The very extension of the notion of space from the material to the virtual has called for explorations of possible relations between the two classes of real and virtual spaces. Shields (2003: xv) referred in this regard to a contemporary "shifting relationship between the virtually real, and the material". Moreover, Crang *et al*. (1999) argued for "the virtual as spatial" (p. 11), and that "virtuality [then] is not just something which operates through and across space. It is at its heart a spatial phenomenon" (pp. 12-3). Thus, cyberspace was recognized to be "hardly immaterial in that it is very much an embodied space" (Dodge 2001: 1). The recent introduction of broadband connectivity, which has permitted permanent access to cyberspace, has turned these observations and assessments even more meaningful, in that cyberspace has become completely integrated into daily activities performed in physical space. Already before the introduction of broadband, cyberspace was interpreted by Bolter and Grusin (1999: 179) as a virtual form of Augé's (2000) *non-places*, which he originally proposed for real spaces, such as airports. Cyberspace may, thus, be viewed also as embedded in physical space: "electronic space is embedded in, and often intertwines with, the physical space and place" (Li *et al*. 2001: 701). Furthermore, both spaces, the physical and the virtual, co-evolve in that they "stand in a state of *recursive interaction*, shaping *each other* in complex ways" (Graham 1998: 174). The power of cyberspace in its constitution of a virtual entity presenting and representing real space has been discussed elsewhere (e.g. Kitchin 1998, Dodge and Kitchin 2001, Kellerman 2002).

The Web, as an application of, or as a rider on the wider entity of cyberspace, may be viewed as a special form of social space. Like the more real social space it constitutes a resource and a production force, for instance for the very existence and functioning of online shopping. Further like social space, Web applications in cyberspace may be looked upon, by their very nature, as invisible texts and as symbols for individuals and organizations, and they may further serve as organizational frameworks notably Intranet systems within work places and multi- location companies (Kellerman 2002). Several commentators claimed for cyberspace *per se* to constitute also a landscape, a place and even a social value (see Dodge and Kitchin 2001 for detailed discussions). Such views substantiate the third definition of cyberspace as conceptual space. Needless to say, though, that in its Web form, cyberspace constitutes an *imagined* space of representation, through its virtual imitation or virtual description of real spaces and places.

Real space and cyberspace have also been viewed as interfolded into each other: cyberspace is accessed from real space, and it contains data on material space, and thus affects it. This relationship has become powerful with the introduction of images of real space by *Google* through *Google Maps* and *Google Earth* for satellite images, permitting users to manipulate the *Google* produced materials and to use them for spatial navigation, thus creating *DigiPlaces*, or blends of the digital and the real (Zook and Graham 2007a, 2007b, 2007c, Crutcher and Zook 2009). Furthermore, cyberspace keeps or imitates various features of physical space. "Virtual environments contain much of the essential spatial information that is utilized by people in real environments" (Péruch *et al.* 2000: 115).

Cyberspace was argued to have its own geography and to be symbol-sustained (Benedikt 1991: 123, 191, Batty 1997). It was further suggested that cyberspace enables and constraints its users like real space (Adams and Ghose 2003). Side-by-side with the connections and convergences between real and virtual spaces, cyberspace is distinguished from real space in many instances (Table 7.3). These differentiations may be divided into three groups of dimensions: organization, movement, and users, and they are discussed in detail elsewhere (see Kellerman 2002: 33-8). It will suffice here to notice that the physical space of virtual space, namely the fixed computer hardware for its operation, may be viewed from a spatial perspective as auxiliary physical space for accessing cyberspace by users. This physical space/hardware seems to have lost its spatial significance once access to virtual space has become fully mobile through mobile phones and portable computers.

Table 7.3 Real and virtual spaces

	Dimension	Real space	Virtual space
Organization			
1.	Content	Physical and informational	Informational
2.	Places	Separated	Converge with local real ones
3.	Form	Abstract or real	Relational
4.	Size	Limited	Unlimited
5.	Construction and maintenance	Expensive and heavily controlled	Reasonably priced and lightly controlled
6.	Space	Territory/Euclidean	Network/logical
7.	Matter	Material/tangible	Immaterial/intangible
Movement			
8.	Medium	Transportation	Telecommunications
9.	Speed	Depends on the mode of transportation	Speed of light, constrained by infrastructure, costs, regulations etc.
10.	Distance	Major constraint	Does not matter mostly
11.	Time	Matters	Matters, but events can suspend in time
12.	Orientation	Matters	Does not matter
Users			
13.	Identity	Defined	Can be independent of identity in real space
14.	Experience	Bodily	Imaginative, metaphorical, close to reality
15.	Interaction	Embodied	Disembodied
16	Attitude	Long-term commitment	Can also be uncommitted
17.	Language	National-domestic	Mainly English-international

Sources: Table: Kellerman 2002: 35, Kellerman 2007, Kellerman 2010b. Items 1-2, 8-12: Li *et al.* 2001; items 3, 14-15: Dodge and Kitchin 2001: 30; 53, items 6-7: Graham 1998.

Navigation and manipulation of cyberspace through the Internet involves a metaphorical spatial experience through the extensive use made of geographical language, symbols and tools, such as homepage, surfing, navigating, site, cursor, etc. The constant discourse between real (material) and virtual (imagined) spaces through knowledge and information (*vis-à-vis* perception and cognition), as well as through spatial experience (real and imagined ones), accentuate the oneness of the material, the perceived and the imagined, as different forms of social space. Prior to the introduction of the Internet, Harvey (1989: 219) noted on the interrelationships between imagined spaces and real ones: "The spaces of representation, [therefore,] have the potential not only to affect representation of space but also to act as a material productive force with respect to spatial practices."

Virtual and real spaces are also interrelated in more indirect ways. For instance, it was shown that persons who navigated successfully through a virtual maze also presented more successful way-finding in real space (Péruch *et al.* 2000). Generally, then, information, knowledge and experience constitute mediating forces between the construction and reshaping of both real and virtual spaces.

An important feature of cyberspace from the spatiality perspective of its users is that it permits simultaneous human co-presence in both physical and virtual places, whereas corporeally a person can only be at one physical place at any given time. Adams (1995) argued concerning the pre-Internet era for human extensibility to constitute a "transcendence of place" (p. 269), and for the media to amount to "extensions of man" (p. 269), notably at times of globalization. Cyberspace and broadband access to it have made this transcendence of place and extension of humans into routine realities.

From a cultural-geographic perspective, cyberspace was argued to amount to a Heavenly New Jerusalem (Benedikt 1991, Wertheim 1999). As such, cyberspace was assumed to move contemporary society from a mere conception of physical space to a more complex one involving also an inner spiritual space, achieved through networks based on global information sharing (Wertheim 1999).

Cyberspace may be divided into two classes in terms of purposes and uses: information [cyber]space (mainly the Web) and communications [cyber] space (mainly e-mail and Web 2.0 applications such as *Facebook* and *Twitter*) (Kellerman 2007). Information space refers to digital information sets or systems, consisting of information organized along spatial notions such as the Web, and, therefore, involving geographical metaphors such as sites, homes, and navigation/ surfing. Information cyberspace further refers to digital information sets at large, such as data archives and library catalogues (Fabrikant and Buttenfield 2001, Couclelis 1998). All these information sets are textual and/or graphic and they have some constancy in terms of their virtual availability to users, so that they may be recalled. Most of these information files are meant to be shared by users: either the general public through the Internet, or segmented and permitted users through Intranets. Contemporary search engines have allowed for easy access to websites and files. *Google* has emerged as a leading service in this regard, providing also searches into specialized information systems, such as satellite images, and scientific articles and books, turning into what one termed as a megaproject within another megaproject (the Internet) (Paradiso 2010).

The second class of cyberspace is communications cyberspace, referring to the cyberspace of persons who communicate via various modes of communications: first, and foremost, through video communications, which includes the transmission of real space visible in the background of the communicating parties, as well as the images of the callers themselves; and second, through purely electronic and invisible spaces of interaction, using non-video communications media (mainly written e-mails, faxes, SMSs, and audio telephone calls) (Kellerman 2007). Communications [cyber] space is mostly interpersonal or shared by small groups, though it may be more widely accessible to larger groups in social networking systems, such as blogs, *Myspace, Facebook, Twitter*, and *Usenets*. Much of the contents of communications cyberspace is not recorded, and if recorded it is meant to be shared only by communicating parties.

The two digital/virtual spaces of information and communications are frequently interfolded, for example when e-mails are sent through an informative website rather than through an e-mail interface, or in messages transmitted through *Usenets*, blogs, *Facebook*, and *MySpace*, which may include pictures and/ or data. Such interfolded and even fused information and communications [cyber] spaces attest to the oneness of the Internet from the usage perspective, being much beyond the shared communications infrastructure of the two spaces. However, each of the two cyberspace classes may frequently function independently of the other, for instance oral personal communications almost always does not involve the transmission of textual datasets.

The very use of cyberspace has involved, until the introduction of broadband for desktop PCs, preparations for the very access of computers and their connection to the Internet, since the billing of Internet use was by the length of use sessions. Following the introduction of broadband in the late-1990s computers have become continuously connected to the Internet by broadband subscribers, as the charge for its use was no longer determined by time of use. The use of the Internet was typified by co-presence of users in both fixed physical location and in virtual mobile spaces (see e.g. Kaufmann 2002: 28, Urry 2000: 71), or by simultaneous embodied and response presences (Knorr-Cetina and Bruegger 2002). However, the introduction of mobile Internet connection through Wi-Fi and cellular modems, has turned the exposure to the Internet instant and permanent, first through laptops and later, as of the early 2000s, also through mobile phones, notably through the so-called *smartphones*. Under such circumstances it is difficult to clearly identify co-presence, or simultaneous embodied and response presences, notably from the perspective of users. On the other hand, Arminen (2007) noted "a dual nature for mobile media, making them both global and local" (p. 432), permitting distant presentation of the self. Furthermore, global contacts may be similar in their social nature to local ones (Knorr-Cetina and Bruegger 2002).

The discussion in this section has attempted to show that cyberspace is simultaneously an entity of its own and an entity converged with real space. The convergence of real and virtual spaces applies mainly to services which are offered in complimentary forms in the two spaces, such as shopping and banking, as well as to cyberspace as representing real space, such as in *Google Maps* and *Google Earth*. The recently growing tendency of cyberspace to converge with real space has emerged through the removal of the distance or the barriers which separated between these two classes of space, in form of cumbersome and costly Internet connectivity and in form of slow and restricted pace and volume of activity. These distance and barriers have been removed with the introduction of broadband services and even more so with the emergence of mobile broadband services. The instant access to cyberspace offered by broadband services has made users take the Internet and its services for granted and see them fully integrated in their daily routines. This applies even stronger for Web 2.0 applications, which call for continuous or very frequent attention by their users, permitted mainly by mobile broadband connectivity, mainly through hand-held smartphones.

The cognition of cyberspace

Following the expositions in the previous sections on cyberspace and its division into information and communications cyberspaces, as well as the discussions on the various communications technologies, we will focus, in this and in the next section, on the cognition of cyberspace and on the instantaneity of mobile broadband services. The basic notions of spatial cognition, spatial perception and mental maps have been reviewed elsewhere (Kellerman 2007), so that our focus here will be on the cognition of information and communications cyberspaces.

Cognitive information cyberspace

We believe that similarly to individuals' cognition of real space they cognize the two classes of cyberspace in special ways. Hochmair and Frank (2001) proposed a metaphorical cognitive mapping for semantic information space consisting of four fields: action affordances for events; physical object hierarchies for substances; attributes for qualities; and user intended actions for activities. Studies focusing on way-finding in virtual reality and studies of Web navigation have both shown some similarity between the Web and virtual reality, on the one hand, and real space, on the other, as far as way-finding and navigation are concerned (see Dodge and Kitchin 2001: 172-3). Still, though, it turned out more difficult for individuals to navigate in cyberspace than in real space, when using virtual navigation technologies available in the mid to late 1990s. As far as the cognition of virtual landscapes is concerned, Internet surfers see only one visually-restricted Web page at a time, which makes it difficult to fully cognize a virtual landscape (Kwan 2001). This limitation may, though, change in the future as a result of further technological developments.

Technologies of Web-search have developed immensely during the 2000s, for both textual and spatial searches and navigations. Textual search is now mostly based on vertical–hierarchical structures, permitting search from general to more specific terms, rather than the simple, and horizontal, space-like searches of specific terms only. Spatial search and navigation through virtual environments permit now the use of directions and angles, as well as looking at virtual landscapes from the perspective of walking persons. Locating addresses on virtual maps and navigating to these addresses has become standard on the Web, as well as in car driving through GPS. In general, as we have noted also for *Google*, textual search and spatial manipulations have been made widely possible and have turned sophisticated.

Cognitive maps for cyberspatial elements such as virtual landscapes, virtual maps, and navigational entities, differ from internal (in mind) and external (on paper) cognitive maps for real space. The basic source of difference between these two types of cognitive maps lies in the basic difference between real space and cyberspace. Real space is experienced bodily and mentally through all the senses. By the same token, cartographic maps of real space are concrete documents and stable entities of information. Cyberspace, on the other hand, is a more flexible and changing mode of information, sensed in restricted ways, normally visually or audio-visually only. Based on Golledge (1995,1999), Kwan (2001: 26) claimed that "without the sense of location, distance, and direction necessary for the formation of configurational spatial knowledge, and without a habitual movement pattern essential for developing route-based spatial knowledge, an articulated cognitive map of cyberspace cannot be established."

Internet users may over time get used to, and thus comprehend, surfing procedures for reaching selected websites or software and then navigate through them, but they may have difficulties to cognize and eventually draw cognitive maps of virtual landscapes or of virtual cartographic maps which they may have been exposed to in restricted sensory ways over the Internet. Furthermore, and as mentioned already, virtual landscapes or maps can be manipulated in varied ways, such as in changing their scale, size, directions, colors, richness of information, etc., and virtual texts too can be manipulated through changing of their formats, fonts, color, etc. Such manipulations may add to the difficulty to cognize cyberspace presentations in memorable ways. As Kwan (2001) noted for real space, space and its maps are two completely separated entities, whereas in cyberspace they may converge.

Cognitive communications cyberspace

Cognitive communications cyberspace might constitute a metaphorical space, notably for the audio telephone, which merely permits a feeling of 'presence' and intimacy by the communicating parties: "'The virtual' is imagined as a 'space' between participants, a computer-generated common ground which is neither actual in its location or coordinates, nor is it merely a conceptual abstraction, for it may be experienced 'as if' lived for given purposes" (Shields 2003: 49). However, cognitive communications cyberspace may constitute more than a purely imagined and metaphorical space, since it may refer to the real spaces of two communicating persons using video media, so that it may be sensed, experienced, or just imagined over the contemporary variety of communications media.

Cognitive communications cyberspace differs widely from face-to-face communications in real space. When two people meet physically, the environment in which their meeting takes place constitutes an element of the meeting, and in several ways. Directly, there are sights, lights, smells, weather conditions, and noises, all of which may draw the attention of the meeting people since all their senses are active. Indirectly, the meeting environment may serve or deter the exchanges between the persons meeting, in terms of the meeting setting and atmosphere. For example, a quiet and cozy restaurant may fit certain meetings more than a crowded fast food facility. All these elements are almost absent from virtual communications, as each party is located in her/his own physical environment and what can be seen and heard by the called party through electronic communications of the surrounding environment is partial and belonging to the separate and unshared physical space of the called party.

Cognitive communications cyberspace also differs widely from the more conventional cognitive space (Table 7.4). Above all, cognitive space evolves and nests in one's mind mainly and routinely for the purposes of orientation or navigation in space (Chang 2003, Passini 1984), whereas cognitive communications cyberspace constitutes a component of interpersonal communications among people. Thus, the elements of each cognitive entity are different. Cognitive space is dominated by physical elements, such as paths, landmarks, etc., while people are not necessarily part of such cognitive space. On the other hand, cognitive communications cyberspace is focused foremost on people, the communicating parties, so that the surrounding physical environment, which may be viewed in video conversations, constitutes background only. On the other hand, however, cognitive communications cyberspace involves several additional elements, pertaining to human contacts. By its very nature, language is essential for communications notably for online spoken or written communications which do not leave time for translation, something which is possible in the use of data or information in information space. The time framework is also essential if communications takes place in real time, and when the two parties are located in different domestic or international time zones. Such time differences apply not only to daily differences but to weekend extents and holiday differences among nations, as well. Less crucial elements are weather conditions, which may potentially deter communications, and international currency differences, if merchandise and services are sold/bought internationally. Altogether, the quality of communications is of much importance, notably for video conversations, which may profit from broadband transmissions probably more than other forms of communications and information transmissions.

Table 7.4 Cognitive space and cognitive communications cyberspace

	Cognitive space	Cognitive communications cyberspace
Objective		
1.	Orientation tool	Communications tool
2.	Physical elements dominating	Physical elements not required
3.	Persons not necessarily included	Mostly only persons included
4.	Time excluded	Time essential (in real-time communications)
5.	Weather irrelevant	Weather relevant
6.	Capital irrelevant	Capital relevant
7.	Language irrelevant	Language essential
Form		
8.	Can be drawn	Cannot be drawn
9.	Aggregative	Non-aggregative
Facilitation		
10.	Facilitating spatial behavior	Facilitating social and economic behavior
11.	Creating movements in places	Creating movement to places

Source: Kellerman 2007.

The two cognitive spaces differ also in their forms. Cognitive space may yield internal cognitive maps in the minds of perceiving persons. These maps, on their part, may be externalized and drawn on paper. For cognitive communications cyberspace, such internal and external mappings seem irrelevant. Furthermore, cognitive mapping of the cyberspace of interpersonal communications may be even less possible than for information cyberspace of the Web, since the spaces of the communicating parties are not exposed to the communicating parties in non-video exchanges, or they may appear only as background in video interpersonal communications. In electronic communications it is for the communicating parties and the contents of the communications session to be in the center of the communications process with less attention given to the physical environments in the background of the communicating parties.

Following Shum's (1990) distinction between locational and attributional information for traditional cognitive maps, we may claim that cyberspace of interpersonal communications, as well as the cognitive cyberspace for such communications, include attributional information with no, or just a little, locational information. Furthermore, cognitive communications cyberspaces are personally unique, and cannot be aggregated, whereas cognitive maps relating to a specific area may be compared and conclusions on a wider societal knowledge of an area drawn. Cognitive space and cognitive mapping may facilitate *spatial behavior*, or corporeal personal mobility, whereas cognitive communications cyberspace may facilitate *social and economic behavior*, in form of interpersonal communications or e-commerce, respectively. Related to this, cognitive space may facilitate navigation or movement *in* places, whereas cognitive communications cyberspace may facilitate movement *to* other places following virtual contacts made with people in them.

Instant mobile access to cyberspace

Following our attempt to understand cyberspace and its cognition through the two components of cyberspace we turn now to the growing use of cyberspace or the growing virtual daily mobilities via the Internet. As we have seen already, the most dramatic recent trend in Internet use has been the development and adoption of mobile broadband devices and services, permitting location- free, instant access and use of the Internet, and "where connectivity functions without regard to place" (Mok *et al.* 2010). Castells (2009: 65) stated that mobile broadband is for the Internet what the electric grid has been for the provision of electric power, i.e. it permits universal distribution. In this section we will first explore data traffic divided by type, followed by a discussion of instant usages of mobile broadband services and their implications (Kellerman 2010b).

Mobile broadband traffic

Data for mobile broadband traffic for the second half of 2009, based on 180 million users worldwide, complement and fortify the adoption trends for mobile broadband connection which we have noted in a previous section (Allot Communications 2010). The breakdown of applications for mobile data usage shows that HTTP (Hypertext Transfer Protocol) streaming, or the transmission of movies and video clips, constitutes the most widely used application (29 percent), followed by HTTP browsing (27 percent), HTTP download (19 percent), P2P (peer-to-peer), or file sharing (19 percent), VoIP and IM (Instant Messaging) (3 percent), and other applications, which would also include e-mail (3 percent). The usage of global mobile bandwidth increased in the second half of 2009 by 72 percent, as compared to the first half of 2009. It was for HTTP streaming to present both the largest share and growth rate within broadband flows, with *YouTube* alone responsible for 10 percent of the global mobile data bandwidth. HTTP streaming increased its share within broadband flows by 50 percent, compared to the first half of 2009, and it grew at that time by 99 percent. Other applications of broadband flows increased significantly as well: VoIP grew by 47 percent, and HTTP downloads grew by 73 percent. Thus, mobile broadband suffered from similar problems to those of fixed mobile broadband: congestion and enormously growing demand.

The fastest-growing region for mobile data usage in the second half of 2009 was Asia and the Pacific, which, as we have seen, was also the region of the widest adoption of mobile broadband subscription, notably in Japan and Korea. For these two countries this leadership continues their pioneering leadership at the time in the adoption of fixed broadband lines (Kellerman 2006b). In Korea, the globally leading country in this regard at the time, some 57.3 percent of Internet households were connected via broadband back in 2001, whereas for the US this rate stood at that time at 11.1 percent. The percentage broadband subscribers out of total Internet users for June 2006 showed that 37.5 percent of Korean users subscribed to fixed broadband, whereas for the US this rate stood at that time at the rather low rate of 27.2 percent. The leading country for the adoption of fixed broadband in mid-2006 seemed to be Japan with 71.4 percent of its Internet users subscribing to broadband, almost twice the Korean (and French) rates.

Instant mobile cyberspace uses

The Internet world consists of three groups of actors: Infrastructure providers; contents providers; and Internet users. The first group consists of providers for telecommunications infrastructures, hosting servers, and Internet and service providers, as major examples. The second group consists mainly of website owners and developers, as well as individuals who serve as contents contributors, in blogs, *Facebook*, *Twitter*, etc. The third group consists of users of two types. The first type includes individuals and businesses that may focus frequently more on Intranet applications, rather than on the Internet, for in-house communications, whereas the second type of users consists of the huge and growing numbers of individual users/subscribers of the internet, who may simultaneously serve also as contents providers for any websites. Our discussion here will focus mainly on this last type of individual users, for whom the mobility revolution of the Internet is of striking significance. Obviously these three groups are interrelated by demand and supply relations, as well as by the complex role of users who may also contribute materials to the Internet, side by side with its consumption. One can also identify an additional, fourth and much smaller group of Internet actors, namely the students of the Internet, whether academic or commercial, whose nature is obviously completely different than those of the three previously mentioned groups who make up the very functioning of the Internet world rather than its study (which may though affect indirectly also Internet uses).

Before moving into a discussion of Internet users it is in place to note, from the perspective of the first two groups, i.e. the providers of infrastructure and the producers of institutional or commercial websites, that cyberspace constitutes for them an entity with specific tools for its operation, maintenance and governance, mostly distinguished from the tools and rules for operation within physical space. Frequently, though, the rules for activities within cyberspace may imitate metaphorically those in real space, such as pressing virtual buttons on a computer screen, buttons which may look like material ones.

For users, cyberspace was, until the introduction of fixed broadband, a completely separate entity from real space entities, as it required logging into the system, involving payment per time use, slow functioning, and limited functionality because of lack of applications. Fixed broadband has changed all this, by permitting constant connection, fast responses and growing numbers of applications for daily uses, such as e-banking, travel reservations, e-government and e-commerce. The introduction and availability of mobile broadband amounted to the availability of these and other Internet services and information without location and time constraints. It has further permitted a wide flexibility of work location, introducing a so-called 'hybrid workspace', though only moderate numbers of workers split their work location between office, home and elsewhere (Halford 2005, see also Hislop and Axtell 2007) (see Chapter 11).

Furthermore, the growing popularity of Web 2.0 applications for social networking has implied a potential permanent presence of users for both consumption and production of Internet materials. Mobile broadband has, thus, made the Internet become a completely routine component of daily life, without any barriers of access or some conceptual distance between it and real space. There is no need any more for users to move to a different arena at a special location (i.e. moving to the location of a desktop PC) in order to access and use the Internet. This change has permitted the development and production of a growing number of applications sold through virtual stores of mobile phone manufacturers. Thus, a vicious cycle of growing use (or demand) and growing supply has emerged.

The omnipresence of the Internet and its instant accessibility has amounted to a practical, rather than theoretical, integration of physical and virtual spaces for users. This instant access to broadband services has some additional implications concerning contemporary society, which is typified by the growing availability and use of communications media. First, growing virtual mobility may increase rather than decrease physical mobility, since one does not have to be tied to a desktop any more in order to instantly initiate, receive and respond to e-mails and other communications, including long international telephone calls via free or low cost VoIP services. Second, the speeding up of daily activities, whether for production or for social communications, may reach now a higher level, since all communications and information media have become fully mobile, thus prompting continuous attention by users. Much before the introduction of the commercial Internet, Virillio (1983: 45) called our era *the age of the accelerator*, and this nature of contemporary society has been accentuated time and again notably regarding car driving, or accelerated and personal physical mobility. "Speed is the premier cultural icon of modern societies…Speed symbolizes manliness, progress, and dynamism" (Freund and Martin 1993: 89, see also Kellerman 2006a). Third, the blurring of separation between work/business and leisure which has typified the spheres of work and home in recent years will intensify, since work and social activities will be easily performed when away from both office and home. Thus, from the perspective of the location of work activities, Castells (2001: 234) referred to *nomadic workers*. The blurring between leisure and work activities is further amplified, since for both activities the same equipment, software and channels are used (see Kellerman 2006a). At the sphere of social relationships, Licoppe (2004) recognized an emerging pattern of continuous 'connected relationships' through various media of electronic communications, so that "the boundaries between absence and presence eventually get blurred" (p. 136).

The use of either mobile phones or wireless Internet connection implies further a blurring between the private and the public, as well as between indoors and outdoors (Kopomaa 2000). Whereas telephones and computers were traditionally considered devices to be used indoors and involving some privacy of communications, wirelessness implies less privacy and a change of social boundaries given the acceptance of communications activity in the public sphere. Thus, the contemporary social environment is characterized by a blurring between the public and the private through the use of mobile phones, into what Sheller (2004b) termed 'mobile publics'. Furthermore, mobile surfing of the Internet accentuates placelessness which may arise through web surfing at large. Virtual mobility via the web implies exposure to remote places while being physically located in a fixed place or while being on the road. "The contradictory experience of being somewhere and nowhere at the same time is perhaps the most obvious cognitive dissonance resulting from the use of the WWW" (Kwan 2001: 26, see also Kellerman 2002: 39-41, 49).

The introduction and wide adoption of mobile broadband implies instant access to cyberspace for its users, and it may also carry implications on the use and meaning of urban physical space (see Chapter 10). The constant availability of GPS, *Google Maps*, and LBS, even while walking or driving through an unknown city or through unknown parts of a known city, implies an efficient moving of people through urban space aiming at specific addresses, thus saving time and efforts in walking, driving, and searching. However, this efficient space crossing turns the crossed streets and urban space at large into a kind of impediment rather than into a cultural occasion for exploration. This specific spatial dimension of the speeding up of mobility turns cities into a mere mosaic of places of production and consumption ignoring the traditional role of cities as providing residents and visitors with passive or active experiences of human life at large, such as city rhythms at different times of the day and the week (see Allen 1999). This loss of meaning of urban space may be tied to a potentially growing loss of contact with the immediate physical environment by mobile broadband users while moving through the urban public sphere, because of their constant engagement in global communications for social networking, services, and information. The term embodiment may, thus, lose its original literal meaning when virtual space is fully integrated with the physical one, and it may be spared for audio or audiovisual exchanges with some partner.

The built-in installation of GPS components in smartphones permits not only access to locational information but also the other way around: the exposure of users' locations to LBS providers who might interfere with the privacy of users, even though they are located in public space (Gordon and deSouza e Silva 2011).

Traditional interpersonal communications has involved strong spatial elements, either through face-to-face meetings in real space, or because it was based on postal communications which carried clear geographical connotations through stamps, the use of street addresses, and the mentioning of writing locations on top of letters. The spatial elements in electronic communications have obviously become weaker, but they have not vanished. The possible *Death of Distance* (Cairncross 1997) in electronic communications does not imply also the potential death of space, the sister primary geographical notion. Electronic communications media mostly do not constitute stand alone forms of communications, so that repeated virtual communications among parties may frequently lead to their face-to-face meetings (see Boden and Molotch 1994, Urry 2002, Tillema *et al*. 2010). Even for distance, the direct significance of which in our lives may seem to have weakened, it was shown by Mok *et al*. (2010) to be still of importance. Comparing communications performance in the same Canadian urban setting for 1978 and 2005 they showed that distance is still significant for communications. The introduction of e-mail in the 1990s has brought about a total increase in communications activity at large, with more e-mailing than face-to-face and telephone contacts. Though e-mailing is insensitive to distance, and thus preferred for long distance communications, Mok *et al*. (2010) were able to show that the significance of distance for face-to-face and telephone contacts has remained unchanged 1978–2005.

Conclusion

The telephone, the Internet and the mobile phone are the three currently available technologies for speedy oral, written, and video communications, with the mobile phone being the most widely adopted one worldwide. The smartphone integrates the mobile phone with the Internet, thus providing location-free access to both communications and information.

'Cyberspace' has been shown to be a rather complex entity consisting of two classes consisting of the information and communications components of the Internet respectively. It has been shown elsewhere that the emergence of cyberspace has involved transitions in the very experiencing of space (see Dodge and Kitchin 2001, Kellerman 2002). However, the experiencing and cognition of cyberspace have themselves become more complex, notably because cyberspace has turned more complex, involving multiple virtual experiences of interactions through both information cyber space and communications cyberspace. Instant access to cyberspace for broadband users has enhanced the integration of cyberspace with real space from the perspective of Internet users. Cyberspace and its various features have not been eliminated, but have rather been incorporated into a complex spatial entity of integrated real and virtual spaces. The 'terminals', namely two or more communicating parties or an Internet user and the websites he/she use, are now at stake rather than the 'roads' or channels connecting them.

We can point also to an increasing indirect significance of distance, in the form of a higher awareness of time differences between our real location and other places and countries with the residents of which we communicate, or the relative global positioning of places and countries. Such awareness of relative global positioning of places may also raise attention to climate/weather differences among places, as potential deterring forces for communications, and depending on the nature of communications with overseas parties some interest may evolve regarding social-cultural and monetary differences among places.

The transition from physically written communications, namely letters, to electronic communications, involves spatial as well as temporal changes, with electronic communications permitting to get in touch with any place instantly. Side-by-side with this change, the use of information cyberspace and communications cyberspace becomes increasingly interfolded. Thus, one may make use of information space through websites, coupled with e-mails exchanged through these same websites which may function simultaneously also as communications space. This interfolded use of cyberspace calls for special attention to spatial cognition as a complex process for virtual spaces.

The growing use of mobile broadband services in a variety and instantly changing physical settings may have some implications for the cognition of physical space while being involved in cyberspace interactions. For example, one may ask whether the growing use of 3G video mobile phones will increase the exposure of video conversation parties to billboard and other road advertising seen in the background. By the same token one may ask whether exposure to background city landscapes in interpersonal communications sessions may serve as triggers for increased tourism. And last but not least, would calling parties in electronic video conversations care more for the appearance of their personal spaces, knowing that these are exposed to called/calling parties?

Numerous cutting edge technologies, as well as the development of technological innovations towards future adoption are aimed at mobile broadband uses. For example, remote access to one's desktop computer, notably to the 'my documents' file; the development of folding screens; 'cloud' Internet storage space; and the development of Tablet computers, equipped with extremely high-resolution screens, encouraging electronic book reading on the go. These and other innovations and applications will encourage further adoption of mobile Internet devices, culminating in an inability to lead personal and household lives without the use of mobile Internet devices. Early trends, notably among younger people, show that the focus of life, notably social life and service provision, may move from physical space to virtual cyberspace as the most obvious arena for their operations. It would be of interest to see, in this regard, whether the natural human ability for spatial navigation will be reduced when extensive use is made of GPS and *Google Maps* while on the go.

For futurists, the current phase of mobile broadband adoption and the blurring of space constitute only another phase of ubiquitous computing towards a full integration of cyberspace *into* material space. Greenfield's (2006) vision of *Everyware* assumes a future disappearance of all information machines such as PCs and telephones as information mediators, since accessing information and having it transferred will be continuously possible through all material entities and in all environments, since information will be disseminated in them, so that the need to get connected to information will disappear as well.

As of early July 2010, Finland, a veteran leading country in telecommunications adoption, has become the first nation to make broadband access a legal right, some 76 years after the US Telecommunications Act (1934) which declared access to telephone service a universal service. Though the Finnish enforced broadband service is rather modest, 1Mbs for downstream traffic, it signifies to service providers, on the one hand, and to the Finnish population as customers, on the other, that broadband services have become a basic form of communication. It would be of interest to see how and when the use of the Internet will become even more popular and wide ranging under such national circumstances. For the opposite end of the 'digital spectrum', the ITU (2010b) reported for 2009 on numerous countries from all world regions to have no broadband subscribers at all. For example: Afghanistan, Central African Republic, Chad, Eritrea, French Guiana, Haiti, Honduras, Kosovo, Liberia, and Yemen. Given the wide implications for individual and societal qualities of life and opportunities associated with broadband, the global digital gap seems to persist, even though basic mobile telephony is now available in 90 percent of African and Asian villages.

Chapter 8
Aerial Business Travel

Following the last two chapters which focused on terrestrial and virtual media for daily mobilities respectively, this chapter will complement this array of mobility media for daily travel by exploring business mobility by air. As we will see, it is rather rare to find commuters who travel daily to their work-place and back home by plane. However, there are numerous workers who travel regularly by air for business and work, and this chapter will, thus, focus on daily air travel in the sense of routine flying for work purposes. Air travel is typified by its being almost solely public transportation operated via airlines. Thus, the deterioration of public mobility services as a result of flourishing personal ones, which we noted with regard to terrestrial and virtual mobilities, is irrelevant for air travel. However, the high cost of air travel as compared to terrestrial and virtual mobilities excludes low income workers from using it on a routine basis unless travel costs are paid for by their employers as it turns out to be the case for many employees. Another difference between terrestrial and virtual travels, on the one hand, and air travel on the other, relates to trip duration which in routine air travel is normally much longer than those in terrestrial and virtual ones. This implies that the conditions and the very experience of being on the road, in this case being on board, are of more significance than those in other types of mobility. Thus, the need of frequently flying business passengers for comfort expressed often in their preference for business class.

Airplanes carry on board business travelers and leisure tourists alike, seated next to each other in both economy and business classes. Hotels, mainly in downtown or in hotel districts of major cities, serve indiscriminately business and leisure guests, which is true also for restaurants and other entertainment establishments. Still, distinctions are normally made, by laymen as well as by professionals, between business travelers or business tourists, on the one hand, and leisure tourists, on the other. Business travel is variously defined as travel or tourism. It refers to people traveling for their work in four classes: individual business travel, meetings, exhibitions, and incentive travel. Of these, individual business travel seems to constitute 'pure' business travel, seemingly distinct from leisure tourism.

The focus of this chapter will be on individual business travel, as a sub-class of business travel, the class which relates mostly to routine travel. The following discussions will provide a distinction between business and leisure travels through their comparison as for motivations and goals, relative magnitude, spatial patterns, and interrelationships between these two types of travel (Kellerman 2010a). The discussions will show that clear-cut differentiations between business and leisure tourisms have blurred: business meetings may yield leisure and *vice versa*; there are shared facilities for both types of tourists; and business activities develop during vacations and *vice versa*. This blurring of differences is similar to the blurring between home and work activities, so that mobility at large evolves as a state of life: at home as well as for work. Urban tourism is a joint geographical context for business and leisure tourisms, since both types of tourists tend to visit cities. More generally, three possible phases of leisure/business relationships will be suggested: spillover, complementarities, and fusion. Alternatively, it will be argued that maybe all three phases operate simultaneously. Before delving into elaborations on these perspectives, business travel has to be defined and business travelers classified, in order to see whether business travel constitutes merely a form of travel or if it constitutes also a distinct form of tourism.

Definition and classification

The contemporary, and rather significant, increase in international leisure tourism has received much attention and extensive treatment in tourism studies as well as in similar fields. However, the no less considerable growth in international business travel, mainly as a result of and as an expression of expanding globalization trends, has gone by with little treatment in relevant literatures, probably with just two recent exceptions (Beaverstock *et al.* 2010, Haynes 2010). This lacuna might hint to one of two contrary options: either business travel is considered similar to leisure tourism, or maybe the other way around, namely that business travel is viewed as a form of work for business persons, a form which does not require special and separate attention, so that it may be viewed as similar to domestic office work, just being performed at a distance from home, and occasionally located in foreign countries.

Business travel has been variously defined as both travel and tourism. Ironically, in a book entitled *Business Travel*, Davidson (2000: 1) provided the following definition: "Business *tourism* is concerned with people *travelling* for purposes which are related to their work…general business travel, meetings, exhibitions, and incentive travel" (italics are by the author). In a later text, Davidson and Cope (2003: 3) noted the confusion between 'travel' and 'tourism' regarding business trips, and they distinguished, therefore, between business travel as individual business travel *versus* business tourism, which refers to business persons going for meetings, exhibitions, and incentive travel. Whereas the latter three classes usually include, *a priori*, some elements of leisure by their very nature, individual business travel seems to constitute 'pure' business travel, involving foremost office meetings and, thus, this class of business travel may seem, at a first glance at least, to constitute a distinct class from leisure tourism. The blurring of boundaries between business and leisure tourism at large has been recently noted also by Lassen (2006) and Faulconbridge and by Beaverstock (2007).

In the past, incoming international passengers were normally asked by passport control agents for the purpose of their visit. Noting the blurring of differences between leisure and business travel and experiencing a growing number of multipurpose visits have brought many countries to stop asking incoming passengers about their visit purposes. Coupled with the almost free moving of Europeans within the EU, without any specific entry documentation, has brought about a reduced and rather constrained availability of comparative data, as we will notice later on.

Given the diversity of business travel, our following conceptual discussions will, thus, focus mainly on a comparison between individual business travel and leisure tourism, attempting to highlight the touristic elements in business travel, and the business-like elements in leisure tourism. Corporate travel of the form of 'general [or individual] business travel', as a sub-class of business travel, will exclude 'travelling workers' (i.e. workers whose very work includes travel, such as pilots), and 'working tourists' (such as travel agents) (see Cohen 1974, Uriely 2001). The emerging class of international work mobility of all kinds, performed through airplanes, was called *aeromobility* by Lassen (2006).

Comparing individual business travel with leisure tourism, it is important to note that air travel *per se* may constitute a routine activity for business persons, and flying business people normally continue to do office work also at their places of destination following air travel, which is of the same type as their office work in their own office (e.g. meetings, e-mailing, etc.). For leisure tourists, on the other hand, both the flight and the touristic activities (e.g. swimming, museum visits, etc.) at their destinations may constitute non-routine activities. Through this differentiation we refer to business air trips as routine domestic and international travel experiences of business people for the purpose of business meetings. This stands in contrast to leisure trips which constitute for tourists foremost an opportunity to be engaged in some non-routine touristic activities at some remote destination, an opportunity which can be materialized only through travel. Taking this difference one step further, we may note that business travel may be considered as constituting means for making business whereas leisure tourism may be viewed as an objective by itself. Lassen (2006) pointed to self-determination as the aspect which traditionally divided between leisure tourism, for which self-determination is highest, and business travel, for which self-determination was considered at its lowest, because employers were assumed to determine the various parameters of business travel for their employees. Lassen (2006) was able to show that this distinction has been blurred, since contemporary employees may share with their employers the determination of their travel, involving potentially both business and pleasure elements. Kesselring and Vogl (2010) believe, though, that employees are under pressure to agree to travel as an integral part of their job. The differences between business and leisure tourisms may lead us to a comparative examination of motivations and goals for these two types of trips, which we will present in the next section.

Daily Spatial Mobilities

Commuters by air

Business travelers were related to so far as people who fly to various places on a regular basis. However, given the focus of this volume on daily spatial mobilities, the most relevant category of business and work travelers by air would have been commuters who fly on a daily basis from their homes to their single office location, whether located domestically or in a neighboring country, returning home every day, similarly to terrestrial commuters. Such air commuters are normally not reported in national statistics, probably because of their negligible numbers, given the high commuting prices by air and the long time required for check-in procedures which we will note later in Chapter 10. A 2001–2002 survey of American long distance transportation patterns, relating to trips longer than 50 miles, showed that only 1.5 percent of commuters of such distances did so by air (BTS 2006). If we add the majority of commuters who travel shorter distances as commuters then the percentage of daily air commuters will be negligible. Probably even the numbers of private plane owners travelling daily from home to office is tiny, given that such business people tend to fly to numerous and changing destinations on a regular basis.

There are, however, other workers who fly regularly and are still termed 'commuters'. For Salt (2010) these "range from weekly commuting for periods of a few months up to one or two years" (p. 111) and include mainly European substitute workers or workers who prefer not to move their residence for family reasons. The 2001–2002 survey of American long distance transportation patterns relating to trips longer than 50 miles each, which we mentioned already, found that generally some 18 percent of business trips were done by air, but distributed among destinations unequally: 52.2 percent of international business travel was by air as compared to 6.5 percent domestic travel (BTS 2006). Wickham and Vecchi (2010) identify commuters as being "all in various roles in customer relations and sales. This is unsurprising, since the rationale for the normal regular journey in the industry is to maintain contact with existing customers" (p. 141) and this despite the availability of virtual mobility media permitting business routine maintenance (see also Haynes 2010). A third, and less clear-cut commuting air travel, is the so called *hybrid workspace* (Haynes 2010), "a working environment in which an individual sometimes works in an embodied organizational space and at other times in other locations, such as at home or on the move" (p. 550), thus implying changing virtual, terrestrial and aerial commuting modes on a daily or routine basis.

Business travel (usually to several places) is expected now by many employers as an integral component of many jobs (Lassen 2010). From the perspectives of business travelers at large, and the perspectives of air commuters in particular, routine business travel as an integral component of work duties requires *mobility competence*, in order "to harmonise travel with the requirements of the job" (Kesselring and Vogl 2010: 149), and in order to cope with possible loss of social rootedness when out of home overnight regularly.

Motivations and goals

Spatial mobilities for business purposes at large may be considered as rather stratified and complex in a globalized world, and they include obviously both virtual and corporeal mobilities. Virtual business mobilities, performed through the telephone and the Internet, may offer partial substitutes for corporeal travel, at least for routine business, such as for sales maintenance and its routine boosting. However, the establishment of new business contacts and contracts, involving the establishment of mutual trust, would normally involve face-to-face meetings (see Boden and Molotch 1994, Urry 2000, Kellerman 2006a), something which Urry (2003b) termed as *meetingness*. Thus, business deals are not necessarily anymore direct and straight forward outcomes of face-to-face meetings only, since they may reflect virtual contacts, as well. As Tani (2005) noted, it is 'head-content' rather than 'headcount' that is of importance in contemporary business contacts.

As we have already noted, Urry (2002) categorized the motivations for individual travel in general around the three elements of people, time and place: potential travelers need to meet other people, attend events in time, or see places (or some combinations among these three elements). Applying this categorization to a comparison between business travel and leisure tourism, we can see that travel for business persons always implies meeting people, but it may less frequently involve visiting sites/places or attending events. Business persons might occasionally visit sites, for instance for the development of a new project, and they might attend events, such as an inauguration of a new service or production line, but when travelling to business destinations they always aim at meeting people, both formally and informally. On the other hand, leisure tourism always implies tourists visiting places, and this is also why tourists normally change their overseas vacation destinations from time to time. In addition to visiting places, for leisure tourists a vacation may or may not optionally involve a desire to meet other people, and/or it may or may not involve attendance of events (Table 8.1).

Table 8.1 Goals of business travel and leisure tourism

Goal	Business travel	Leisure tourism
Seeing places	Infrequently	Always
Meeting people	Always	Sometimes
Attending events	Sometimes	Sometimes

Source: Kellerman 2010a, see also Urry 2002.

Relative magnitude

How many international business travelers and leisure tourists are there? This question may be asked for all the three relevant geographical scales: global, national and local. For global and national measures, data on the number of tourists are published by WTO (World Tourism Organization), and these may be helpful for the calculation of a percentage ratio of leisure to business tourists, in order to assess the relative magnitude of these two classes of tourism. Such a simple ratio is, thus, sensitive to changes in the numbers of both leisure and business tourists. However, for measuring the even more intriguing shares of leisure and business tourists at the city level, only the percentage of business tourists has been available, and even this measure could be found for only a handful of cities for which such data have been published, normally by city governments or agencies.

There seems to exist some problem of data reliability shared by all statistics at all the three geographical levels, since WTO collects and publishes national data produced by relevant national authorities, whereas city data may constitute only estimates made by municipalities or other local agencies. As we noted already, it turns out that the traditional classification of tourists into business and leisure has no longer been provided in recent years by several countries, given the complexity of travel motivation, which mixes business with pleasure within same trips.

At the global level, business visitors were variously estimated to range from some 15–20 percent of total visitors (Law 2002) to some 25–50 percent (Haynes 2010). The WTO data show that the global leisure tourists outnumbered business ones by 315.6 percent in 2002; 314.6 percent in 2003; and 321.1 percent in 2004. These ratio levels seem to remain constant in recent years so that globally there are three leisure tourists for every business visitor, or that, in other words, some 25 percent of international travelers are recognized as business travelers. The constancy of this ratio may mean that the global growth in leisure tourism has been coupled with a similar growth in business tourism. However, the possible interdependence between the two classes of tourism is only partial, as we will see later on. Interestingly enough, the most growing world tourism class has been neither leisure nor business tourists, but rather the third class of travelers, namely that of 'VFR (visiting friends and relatives), health, religion, and other'! The highest growth levels reached by this class might be related to the growth in immigration, bringing about more family visits by immigrants in their old motherlands, side by side with visits to the new countries of residence by family and friends of the immigrants who still live in the country of origin.

The relative magnitude of business tourism at the national level is presented here through the ratio of leisure to business tourists 2001–2005 (in percent) for various countries (Table 8.2). The missing of data for leading countries in leisure tourism, such as Austria, France, Germany, Greece, The Netherlands, and Switzerland, stems from the lack of differentiation between leisure and business visitors in their national statistics, presenting, as we mentioned already, a difficulty in distinguishing between the two classes, when visitors enter countries for joint visits involving both business and leisure. It further presents a lack of interest by countries in maintaining such a differentiation when both forms of tourism use similar infrastructures (such as transportation) and services (such as hotels). As we noted already, this tendency of non-differentiation between business and leisure visitors is typical to European countries, since no specific entry forms are required for EU residents moving among EU countries.

Table 8.2 Ratio of leisure to business tourists 2001–2005 (in percent)

Country	2001	2002	2003	2004	2005
Australia	296.8	275.0	271.2	268.8	267.3
Austria	—	—	—	—	—
Belgium	163.2	174.3	189.6	199.4	185.2
Canada	446.4	449.3	433.4	248.3	379.5
China	—	172.7	148.4	192.0	203.2
France	502.1	—	—	—	—
Germany	—	—	—	—	—
Greece	—	—	—	—	—
Hong Kong	—	—	—	—	—
Italy	275.2	259.2	280.4	309.7	282.8
Japan	219.8	240.5	238.5	277.7	295.8
Korea	1769.5	1708.4	1392.6	1567.4	1429.9
Netherlands	—	—	—	—	—
New Zealand	400.4	426.2	413.4	392.7	373.7
Singapore	217.9	226.4	208.9	—	—
Spain	1154.1	1057.3	710.4	782.8	739.0
Sweden	114.6	144.8	140.4	—	—
Switzerland	—	—	—	—	—
Taiwan	111.7	110.3	89.2	103.1	133.8
United Kingdom	102.7	98.7	105.0	115.1	109.9
United States	149.0	141.3	161.5	184.4	181.5
World		315.6	314.6	321.1	

Sources: Table: Kellerman 2010a. Data: UNWTO 2007a, 2007b.

Though the data set used here is partial, it still permits some interesting observations and interpretations. Generally, the countries for which data were available, present stability in the relationship between the two classes of visitors, so that the global stability in this relationship, which we mentioned before, reflects wide national ratio stabilities. This simultaneous growth in business and leisure visits attests once again to a basic element of globalization, that when international movements at large are facilitated and enhanced then such policies would affect all types of transactions and human movements (see e.g. Appadurai 1990, Kellerman 1990, Kulendran and Wilson 2000, Kulendran and Witt 2003).

Most countries enjoy a larger number of incoming leisure visitors than business ones, which is true even for countries which may be called 'business states', such as Singapore. One cannot discern a general relationship between the level of popularity of countries as leisure destinations and their ratio values, so that the ratio levels rather attest to domestic trends. For example, Spain, which is a leading country in leisure tourism, presented lower ratio scores than Korea which is considered more of a business destination. By the same token, both the UK and Taiwan reached ratio levels attesting to an equal number of leisure and business tourists, despite a major difference in their level of attraction for leisure tourists, with the UK being a popular leisure destination, as compared to the more business-oriented nature of most visits to Taiwan. Furthermore, similar ratio values have also been shown by the US and Sweden, probably excluding visitors from Canada and Mexico to the US, and possibly attesting to an insensitivity of the leisure/business tourism ratio to country size. Some other countries exhibit a special business status. Thus, the low ratio values of Belgium, which hosts the EU headquarters in Brussels, and hence attracts many business tourists, and the higher levels of Australia as compared to those of New Zealand, attesting to a higher business attractiveness of Australia as compared to New Zealand.

As it turns out, rarely do cities collect or release data on the categorization of international visitors, so that data could be found through the Web for only five major world cities, albeit in four continents (Table 8.3). The range of the percentage value of business tourists among these five cities is high, ranging from 28 percent in London to 66 percent in Frankfurt. Thus, these rates may be related to local business specialties as well as to the degree of leisure attractiveness of cities. For instance, the high value of Frankfurt attests to its financial centrality, side by side with its lower leisure attractiveness, whereas Boston enjoys visits by some special business communities through its leading health services and universities.

Table 8.3 Percentage business visitors and expenditure among international tourists in selected cities

City	Percentage visitors	Percentage expenditure	Year
Boston	46	n/a	2000
Frankfurt	66	n/a	2006
London	25[1]	33	2003
Singapore	28	35	2006
Sydney	35.5	n/a	2005

Note: [1] Percentage of overnight visitors.

Sources: Table: Kellerman 2010a. Data: Boston: New York, Boston, Washington DC-Media Kit Request 2007. Frankfurt: MPI 2007. London: City of Westminster 2006. Singapore: Newscentral 24 2007. Sydney: Sydney Media 2007.

In many cases, however, and similarly to nations, cities may be attractive to both business and leisure visits. Thus, Faulconbridge and Beaverstock (2007) were able to show that the leading European cities for business visits were London, Paris, Frankfurt and Geneva, and three out of these four cities are simultaneously also major leisure-tourism havens. Still, however, leisure-attractive cities may present varied rates of business visits as compared to their respective national values. For example, the ratio of business to leisure visitors into London is 1:4, as compared to the UK national ratio of 1:1, attesting to the high attractiveness of London to leisure tourism as compared to the rest of the country. In Sydney, the local ratio is similar to the national one, 1:3, which is true also for Singapore in which the two values are similar, given the nature of the country as a city-state (1:3/1:2). London was ranked much higher than Sydney and Singapore in 2006, as far as the populations of the urban areas of these cities were concerned: London was ranked 28 (with a population of 7.61 million), as compared to Singapore's rank of 55 (4.47 million), and Sydney's similar rank of 57 (4.45 million) (Citymayors Statistics 2008). This difference might add to the explanation of the differences among the ratios of business/leisure tourists for these cities, so that a larger city in a larger nation might attract more international leisure tourists. As we noted earlier, possibly the ratio of business to leisure tourists at the national level is less sensitive to population size, but it seems that as far as cities are concerned their size matters, assuming that larger cities provide more tourist attractions of various sorts.

For the two cities for which data on percentage expenditure of business visitors out of total visitors' expenditures were available, business travel expenditures were higher than the percentage of business visitors out of the total number of visitors. This additional contribution to local incomes from business tourism looks as if business tourism 'cross-subsidizes' the development and maintenance of tourism infrastructures for leisure tourism, though it does not necessarily has to be so generally, as leisure tourism may be profitable by its own operation.

Spatial patterns

Spatial touristic patterns present mixed and complex trends when it comes to a comparison between business and leisure tourisms. Normally leisure tourists make use of business infrastructure and services but usually not *vice versa*. Thus, business tourism infrastructures (i.e. hotels, restaurants, etc.) in cities may be useful for leisure tourism as well. In addition, business services for travelers have become important for leisure tourism as well, for example hotel 'business centers', providing access to the Internet, and airline business class, providing higher flight comfort, both of which may be used by leisure tourists, side by side with business visitors. On the other hand, however, infrastructures geared for leisure tourism may be of less or no significance for business travelers. For instance, beaches, historical sites and museums, which serve as major leisure activities and attractions, are normally not visited by 'pure' business travelers.

Preferred locations for business hotels within metropolitan areas have expanded, so that the coincidence of the locations of business hotels and those of leisure-oriented ones is not restricted anymore to downtowns only. Business-oriented hotels have expanded their locations to outlying, suburban and even exurban business centers or *edge cities*, and these outlying areas might coincide with leisure areas and touristic attractions (see Garreau 1991). Business hotels have also been built next or even within major airports, providing diversified amenities for traveler while on the road, and obviously these hotels are also attractive to leisure passengers on transit.

Even more striking are transitions in the specialties of cities, as far as incoming tourism is concerned. Cities which originally functioned as purely business cities have become also leisure oriented ones, for example UAE cities such as Doha, Dubai, and Abu Dhabi. Such a transition may also happen the other way around, so that cities which originally functioned as leisure cities have turned into major business centers, as well, such as Orlando and Las Vegas becoming centers of high-tech R&D (research and development) and production. However, whereas in the Gulf countries the locations of business and pleasure activity areas within cities are mixed, in American cities business areas and their hotels may develop in separate areas detached from the previously developed tourist attraction areas in such cities as Orlando and Las Vegas.

Interrelationships with leisure tourism

The blurring of distinction between business and leisure tourisms is not only spatial but it relates also to the travelers themselves (see Lassen 2010). An interesting example is the attitude and use of time by travelers which presents some interesting interrelationship between business and leisure tourists. Thus, the business accent on time as a major resource has become important in leisure as well, as leisure tourists plan the time frames for their various vacation activities, being aware of the rather restricted time availability for their vacation, and being used to efficient time use from their daily time use at work. The use of business services by leisure tourists, mainly hotel business centers and smartphones or laptops, may facilitate a more efficient time use when on vacation. On the other hand, the accent on pleasure in leisure tourism has taken on growing importance for business, as well, so that relaxed dining has turned into an integral part of business-making itself, and this may apply also to other forms of entertainment, depending more on local business cultures.

The complex interrelationships between the two classes of business and leisure travels come into expression through two special groups of visitors: *returners* and *extenders*. Returners are leisure tourists who happen to like their vacation destination and who decide to return there for the exploration of some business opportunities (Davidson and Cope 2003: 261). Extenders are visitors who *a priori* conduct multipurpose travel to a specific destination, involving both business and pleasure (Davidson and Cope 2003: 257). Studying exports and imports in an Australian longitudinal context it has been shown that levels of international trade were related to international travel at large. Furthermore, it was argued in that study that leisure tourism may bring about business tourism and *vice versa* (Kulendran and Wilson 2000, Kulendran and Witt 2003).

Conclusion

This chapter focused on individual business travel, as a sub-class of business travel. Such travel may constitute a routine mobility activity for numerous workers and business persons, involving the use of planes, hotels and other facilities and services which serve also leisure tourists. As we have noticed, any clear-cut differentiations between business and leisure tourisms have blurred for all the three major dimensions of tourism: people; places and activities. For people, business meetings by business people may yield leisure visits by these business persons and *vice versa*, leisure visits may bring about business ideas and opportunities yielding future business visits by vacationers. As for places and activities, leisure tourists and business visitors may share the same transportation, lodging and entertainment facilities. This blurring of boundaries between business and pleasure tourisms is similar to the emerging blurring of distinctions between daily home and work activities, so that mobility at large, domestically and internationally, routine and non-routine alike, evolves as a rather continuous and permanent state of life involving in a rather integrated way both business and leisure travels (see Urry 2000, Kellerman 2006a). Obviously this integration may also imply contradiction and conflict between business and pleasure, but these aspects are beyond our focus of discussions.

Cities, notably major ones, serve as the joint spatial platforms and meeting arenas for business and leisure tourisms and tourists, and hence the importance of urban tourism. As a starting point for an examination of cities as joint platforms for business and leisure tourism, one could assert that there might possibly emerge three phases of leisure/business relationships in cities. The first phase could be called spillover, in which business or leisure tourisms as a well-established class of tourism in a certain city makes room for the use of its facilities by the other class through the offering of available services; for example, business hotels which become available also for leisure tourists. In the second phase, both types of tourism in a certain city may have become well-established and sizeable, and the two classes complement each other in the creation of demand for additional infrastructures, such as expanded airports, roads and railways. In the third phase, the two categories of tourism may fuse into each other, and it becomes difficult to separate hotels and other urban touristic services and infrastructures by their served markets. Alternatively, it may also occur that all three phases, or forms of relationships, may operate simultaneously within one city, so that there might exist some spillover effects between the two categories of tourism, side by side with complementarities and fusions between them taking place. Each form or phase of relationship may, thus, take place in a distinguished type of urban services or facilities which serve tourists. For example, business hotels becoming available also for leisure tourists, simultaneously with the emergence of demand by the two forms of tourism, for additional infrastructures, such as expanded airports, roads and railways.

PART III
Spatial Implications

Chapter 9
Urban Spatial Reorganization

This chapter moves us into the third and final part of the book. Following the exposure of the roots and nature of daily spatial mobilities in Part I, and following the discussion of the three major types of daily mobilities: terrestrial, virtual and aerial in Part II, this part will have us explore spatial elements of daily spatial mobilities, focusing on possible transitions in the spatial organization of cities, on mobility terminals and on spatial opportunities for mobile individuals.

This chapter will highlight urban spatial organization and landscapes, in lieu of ubiquitous personal uses of ICTs (Kellerman 2009a). It will relate to two classical notions of urban geography and urban studies at large, namely *spatial organization* and *urban landscapes*, in lieu of the contemporary ubiquitous personal adoption and uses of media for virtual mobility. It will present the uniqueness of contemporary ICTs as compared to previously adopted technologies for personal corporeal and virtual mobilities: they do not require ample urban space for their adoption and usage, nor do they practically facilitate new waves of urban spatial growth as the car and telephone did in the past. They rather permit sophisticated uses of existing urban spaces, without regard of specific spatial organizations of cities. As such, urban spatial organization may have reached a saturation level as far as transitions being facilitated and brought about by mobility technologies is concerned.

The chapter will first examine contemporary media for personal virtual mobility through three major facets of urban landscapes: motorized traffic; pedestrians and drivers; and spatial organization. The discussions of spatial reorganization and urban landscapes under the predominance of personal mobilities in the information age will begin with a discussion of urban facets of ICTs for personal mobilities, focusing, first, on mobile artifact motorist traffic, followed by a discussion of people's movements about cities as *homines viatores* (i.e. mobile people). Then, a more elaborated discussion will be devoted to a third urban facet of personal mobilities, namely the fixed artifact environment, or the spatial reorganization of cities in the information age.

Urban facets of ICTs for personal mobilities

As we have seen already in Chapter 7 mobile telephony is the most widely adopted personal mobility device, and its growing integration with Internet communications through smartphones signals a completion of a major phase in mobile information society, permitting one to have immediate access to all potentially available personal and public information, and permitting constant and universal communications in all possible forms. A contemporary urbanite, notably in developed countries, may now have access and he/she can be accessible to all possible classes of information, whether a person is sedentary, at home or at work, or while being corporeally mobile, on the street, or while stopping somewhere, e.g. in stores, or in cafés. Such universal communication and information access is normally achieved through one or two small devices, the mobile phones (mainly smartphones) and Tablets or laptops.

A device/technology of particular significance for urban terrestrial mobility is GPS, demonstrating another facet of terrestrial and virtual mobilities. GPS devices are satellite communications devices, usually mounted in cars and integrated into smartphones, providing vocal and visual navigation assistance. They permit easy, safe and direct navigation, especially when driving through unknown urban environments. The penetration of GPS is still in its early phases as compared to earlier devices. The US is one of the three leading countries in GPS penetration (jointly with Japan and Korea), with some 21.5 percent of its adult population using GPS devices in 2009 (GPS Magazine 2009).

The examination of ICTs for personal mobilities in cities involves three major and interrelated facets of urban landscapes. First, *mobile artifact motorist traffic*, or mechanized terrestrial mobility throughout the city, including cars, motorcycles and bicycles for personal corporeal mobility, side by side with buses, taxis and light trains for public physical mobility. Second are people in cities, including residents, commuters, visitors, and business persons (see Martinott 1994), all producing pedestrian traffic, alongside persons who drive vehicles or ride in them, all of whom we will term later on as *homines viatores*. Third, is the *fixed artifact environment*, consisting of urban buildings, streets, and other land-uses. The major question to be discussed is, thus, whether the ubiquitous adoption and use of mobile devices for virtual mobility, coupled with the massive penetration of the Internet as an information machine, has changed these three facets of urban landscapes.

Urban traffic in light of the vast adoption of ICTs has been reviewed and interpreted by Kwan (2007), whereas pedestrians and drivers in contemporary ICT-driven cities have received several treatments (Kellerman 2006a, Solnit 2000, Sheller 2004a, Sheller and Urry 2000). However, the third facet of urban landscapes, that of the fixed environment, has received much less attention. Thus, following brief discussions of the first two facets, our attention will be devoted to the third one. The discussions of the first two facets which focus on traffic and people as the mobile dimensions of cities are important in order to understand the third facet which focuses on the rather fixed landscapes of cities. As we will see, the fixed and the mobile facets of cities have been interrelated by their very nature, and this interrelationship is of even higher importance in the contemporary age of growing personal corporeal and virtual mobilities (Kellerman 2006a).

Mobile motorist traffic

Physically, traffic in contemporary cities does not look very different from the way it looked in the pre-information age, let's say some 30–40 years ago: major streets are flooded with the same kinds of vehicles, including cars, motorcycles, bicycles, taxis, buses and trucks, side-by-side with trams and underground urban railway systems. Most road junctions continue to be managed and controlled by traffic lights, whereas some others have become interchanges consisting of over- and under-passes. The only seeming ICT-based change in mobile motorist traffic seems to be the remote control of traffic lights by traffic control centers using cameras and sensors mounted at road junctions. However, an assessment of contemporary information age urban traffic from the perspective of traffic only might be misleading, even if we can recognize some relaxation of rush hour traffic in some countries, such as France, attributed to flexitime travel based on the use of mobile phones which permit instantaneous changes in travel routing (de Gournay 2002, see also Kwan 2007). Traffic has changed not in its physical appearance in city streets but rather in its constitution of a material expression, or in its embodiment, of changing aggregate patterns and substances of personal mobilities. In other words, urban traffic presents the aggregate ways through which individual drivers and passengers direct and manage the three major dimensions of each of their trips: objectives; points of origin and destination; and routing (Figure 9.1). These three major trip dimensions have changed in the information age in that they all have become much more flexible through the use of ICTs.

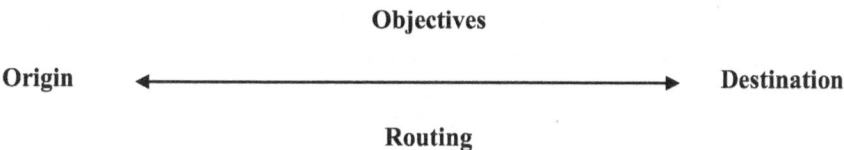

Objectives

Origin ⟷ **Destination**

Routing

Figure 9.1 Major trip dimensions

Trip objectives may change when ICTs are widely adopted, since trips for services which can now be replaced by virtual ones through the Internet (e.g. banking, shopping or browsing, and travel reservations) can be exchanged for other, mainly social, trips (see e.g. Mokhtarian 2000).

Trip origins can be changed due to the blurring of boundaries between work and leisure caused by the use of computers and computerized communications, mainly through the Internet. Thus, one may leave home at times other than the morning rush hour, so that some work, at least occasionally, may be performed from home. Another possibility is for people to leave their work places late, while taking care of some home affairs at work through communications media. Work can further be pursued while being at certain entertainment facilities, such as cafés (Kwan 2007).

Destinations have changed through the adoption of personal virtual mobility media mainly because of the Internet which permits the performance of many activities and services from home or work rather than their being carried out through physical visits to service locations. In a certain way, even the sedentary nature of the notions of 'origin' and 'destination' for both single-purpose and multi-purpose trips has changed since many activities can now be handled while driving, through the use of mobile phones, and an even wider array of activities may be carried out using the Internet when riding public transportation or while sharing carpools.

Trip routing too has become much more flexible in the information age with implications on the final destinations of trips as well. GPS can assist in choosing proper routes providing preference for shorter and less congested ones. The ubiquitous use of mobile phones permits what Ling (2004) termed 'interaction-based coordination', involving three *en route* scenarios of time-space flexibilities: destination changes; destination and schedule setting; and schedule changes (see also Kopomaa 2000, Kwan 2007).

Homines viatores

As the discussion in the previous paragraphs has shown, it is impossible to grasp the changes in urban traffic if only vehicles rather than vehicles and their drivers and passengers are looked at. However, contemporary urbanites should not be looked at as merely engaged in corporeal mobilities when they drive or when they are passengers located in public or semi-public domains. City residents or visitors should rather be viewed foremost as people involved in virtual, ICT-driven mobilities, possibly being simultaneously involved also in corporeal mobility. The metaphor of *homo viator*, i.e. mobile Man, originally proposed by Eyerman and Löfgren (1995) for human life at large, may be useful for contemporary mobile urbanites, as well (see also Chapter 1 and Kellerman 2006a).

Urbanites could be classified, from a communications perspective, as being at any point in time engaged in some form of co-presence. One such co-presence could be termed as passive presence, referring to a condition of somebody being in a public or semi-public domain but not being engaged in face-to-face conversation with others. Another co-presence form is active co-presence which implies involvement in interaction with one or more physically present counterparts. A third type of co-presence is the virtual one, which until recently could be achieved only while being in fixed locations through the fixed-line telephone. More recently, the contemporary urban landscape increasingly presents people involved in a new condition of public virtual co-presence, when persons are engaged in telephone calls, either through mobile phones or through VoIP, or when they are connected to any website through the Internet while being in the public sphere. Such virtual co-presence occurs side by side with the very existence of numerous passive and active co-present people in urban public spaces. The urban co-presence landscape has become diversified and complex. Thus, more and more people can be seen engaged in telephone calls while walking on the streets, standing in line, riding public transportation or driving their cars. More and more people can be seen using the Internet through laptops or through I-mode or 3G mobile phones while sitting in a park or in a café, using Wi-Fi or UMTS connectivity. Other people may be seen occasionally interrupting a face-to-face conversation because of incoming phone calls. The urban public sphere has, thus, become dominated by an urban landscape of much more complex human co-presence patterns when media for virtual personal mobility are widely adopted and used. Urry (2007: 176) termed this scene of communications 'connectivity', emphasizing that people tend to get engaged in the use of mobile communications media while involved in corporeal mobility through walking, riding or driving, or while waiting for a public transportation medium.

Driving personal media for corporeal mobility about the city, whether these are cars, or cycles of any kind, implies continuous impacts by these drivers on fellow street users, drivers as well as pedestrians. Such impacts by drivers of mobility media are, for instance, changes in traffic speeds, lengthening of stops at traffic lights, and stopping for pedestrians when they cross streets. On the other hand, however, the virtual co-presence or mobility of urbanites does not automatically interfere with or having an influence in any way on the virtual co-presences of others who share the same open area, unless there is some communications traffic congestion due to restricted wireless bandwidth. In addition, virtual co-presence can be controlled for the minimization or cancellation of any nuisances caused by users to fellow users of the public or semi-public urban sphere, by using low voice in telephone conversations. Furthermore, one can still assume that most of people's virtual mobility is performed in the private sphere, whereas corporeal mobility can take place only in the public sphere.

People's usage of media for virtual mobility does not only imply that they are connected with other people or with remote information sources. Sometimes, they may be connected with themselves only, using their mobile communications devices for browsing, and updating or manipulating their personal calendars and personal contact lists, or using their built-in calculators, etc. Such organizational and personal management activities may frequently yield, though, immediate new movements or changes in movements in which they are involved at that time, corporeal and virtual alike.

As we have noted in previous chapters, contemporary human communications abilities are typified by several traits: constant potential virtual communications; increased time-space flexibility for corporeal mobility; and a blurring of boundaries between work and leisure (see also Kellerman 2006). These traits may bring about an accelerated pace of activities about the city and beyond it. Such an accelerated pace may not necessarily bring about faster walking and driving paces, because these are limited by the width of roads and sidewalks, as well as by the number of cars and pedestrians, respectively. However, accelerated paces of activity may find their expression in other elements of contemporary lifestyles, for example in the popularity of fast food worldwide.

The urban fixed environment

Urban spatial reorganization was defined already back in 1969 as a process which was brought about by mobility changes: "a process by which places adapt both the locational structure and the characteristics of their social, economic, and political activities to changes in time-space connectivity (the time required to travel between desired origins and destinations)" (Janelle 1969: 348). The question is whether this process of urban spatial reorganization may still be at work under contemporary mobility patterns in which transitions in connectivity have emerged mostly in virtual connectivity rather than in corporeal one.

The assessment of potential changes in the fixed artifact environment of cities in light of the wide adoption of mobile ICTs, or the potential for urban spatial reorganization in the information age, requires a historical comparison of contemporary ICTs for personal mobility with previous technologies for corporeal and virtual mobilities. This comparison has to be performed from a double perspective: first, the requirements for urban space posed by mobility technologies at various historical ages for their very functioning, and, second, the facilitation of urban spatial expansion offered by these technologies along time (Table 9.1). We suggest a distinction among five layers, or historical periods, of urban mobility technologies which will be outlined in the following paragraphs: pre-modern; modern I; modern II; post-modern I; and post-modern II.

Table 9.1 Layers of urban mobility technologies

Layer	Technology	Spatial facilitation	Space consumption
Pre-modern	Horse and cart	Restricted	Relatively minimal
Modern I	Trains	Growth along lines and serving areas	Extensive: Stations, stops, and lines
Modern II	Cars and analog telephony	Extensive growth and serving points	Enormous: Roads, interchanges, parking, garages, exchanges
Post-modern I	Telnet and digital telephony	Edge cities	No spatial requirements other than for computers and terminals getting smaller
Post-modern II	Internet and mobile telephony	Efficient and sophisticated use of any existing areas	No spatial requirements or adjustments

Source: Kellerman 2009a.

Thrift (2004b) presented a classification of the infrastructures for urban mobilities into two waves, which will be outlined below. These waves have emerged within what he termed as 'the surface on which life floats' (p. 584). This surface is the 'first nature', constituting the physical or natural 'background' of 'natural order', or what we may simply call 'nature'. On top of this basic nature, two waves of a 'second nature' have evolved. The first wave of the 'second nature' includes roads, pipelines, lighting, etc., providing for human corporeal mobility, as well as the mobility of electricity, gas and fluids (water and sewage). This first wave of the 'second nature' is meant to constitute and to function like a breathing system for urban systems. Contemporarily, 'a second wave of second nature' has emerged, consisting, among other things, of cables, wireless signals, and artificial fibers providing for virtual mobility by the transmission of signals. Mobility infrastructures consisting of various technologies and generations are, thus, considered as an urban human-made nature. We propose below five layers of mobility technologies in cities which both function within and bring about the construction of the two waves of the 'second nature'.

Traditional pre-modern cities were based on carts and horses for physical mobility and they lacked any technology for personal virtual mobility. Such cities did not need to allocate much space for mobility infrastructures, and simultaneously these traditional cities could not enjoy substantial urban spatial growth when mobility was based merely on walking and animal and carriage riding and use. The advent of mobility technologies until the introduction of current ICTs implied tremendous urban spatial growth coupled with a growing need to allocate significant space chunks for the very operation of these mobility technologies. Thus, the first modern era of the mid to late nineteenth century introduced trains of various kinds, requiring space for lines and stations and permitting urban expansion along railway lines. In the second modern era, stretching through the first half of the twentieth century in North America and through the post-World War II decades in Europe, cars and telephones were massively adopted (see Kellerman 2006a). This adoption implied that much space had to be allocated for the large number of private cars, which required roads, parking areas and maintenance facilities, side by side with large buildings which were constructed for analog telephone exchanges. Simultaneously with these extensive spatial needs of modern mobility technologies, these technologies permitted the vast suburbanization typifying contemporary metropolitan areas through the provision of infrastructures for fast corporeal and virtual mobilities.

The first post-modern era, as far as mobility technologies are concerned, began in the 1980s with the introduction and adoption of digital telephone systems requiring little space for telephone exchanges and permitting the use of advanced computer communications services channeled through the telephone system. This advance was coupled with the introduction of fast and wider bandwidth for data transmission through satellites and fiber optics. All these new technologies permitted the connection of mainframe large computers to remote terminals, thus bringing about a massive suburbanization, this time mainly of office services and company headquarters, bringing about the emergence of suburban and exurban business centers which Garreau (1991) called *edge cities*.

The introduction and widespread adoption of the Internet and mobile telephone technologies and devices from the 1990s has marked the second post-modern era typified by a major change in the relationships between mobility technologies and urban spatial reorganization, in both space consumption of these technologies and in their facilitation of urban spatial growth. Simply put: *contemporary ICTs, such as the mobile phone, the Internet and GPS, do not require any significant pieces of land for their own operations and at the same time they do not facilitate additional urban growth which could not be facilitated by previous technologies.* Potentially, these technologies could have brought about the complete dismantling of cities since access to a full range of information and communications services is now permitted from any point in space. Thus, there were some who predicted at the time such a future disappearance of cities (e.g. Toffler 1980), but these prophecies have not been materialized. It has rather turned out that people need cities for both social and economic aggregations, in what Duranton (1999) termed the 'tyranny of proximity' (see Chapter 2). Of major importance in this regard has been one specific urban section, the CBD (Central Business District) which was weakened at the time, mainly in North American cities, due to the wide adoption of the car and the telephone which have facilitated a suburbanization of businesses. The new ICTs have not contributed to a continued decline of CBDs.

Hence, it is now for the first time in the modern history of mobility technologies that technologies may permit urban spatial expansion at almost no spatial expense for the sake of the technologies themselves. However, a vast spatial growth does not occur, due to social limits. In other words, urban spatial expansion has reached saturation level without regard to the additional spatial expansion which may potentially be facilitated by new mobility technologies.

What then have these new technologies done to urban spatial organization? Not much as far as spatial restructuring or spatial expansion is concerned. Another important contemporary trend, globalization, has also not yet brought about significant change in urban spatial organization, since globalization has implied the connection of people and places separated by long distances and oceans without requiring or necessarily bringing about any transitions in the spatial organization of the connected cities in which they live and work. In terms of land-uses, though, some new facilities have been added to cities, directly devoted to the use of ICTs. These include Internet cafés, and stores and service stations dealing with the purchase and service of ICT devices, mainly computer stores and mobile telephone stores and service centers. In addition, data storage houses and command and control functions for the Internet are new facilities which were added to urban land-uses, but they are of less direct attachment to ICT end-users. Though the consumption dimension of ICTs has not added much or changed much in urban spatial organization, the production end of ICTs has brought into numerous cities high-tech R&D and production parks (see Kellerman 2002, Zook 2005).

As far as changes in land-uses are concerned, directly or indirectly related to ICTs, these have been concentrated in one specific public domain, as well as in almost all private and semi-private spheres of life. The specific public domain of ICT related land-use changes is airports, mainly international ones, devoted to long distance trips, as compared to automobiles which are used mainly for local and regional daily travel (see the following chapter and Kellerman 2008). Growing air travel has brought about the expansion and construction of new terminals, as well as the expansion of airports at large. One of the sources for the tremendous growth in air traffic for both pleasure and business tourism has been interrelated with the growing exposure of people to other countries on a global scale through the Internet and cable-TV.

The private and semi-private domains which have changed through the growing adoption and uses of ICTs are homes and offices in developed countries, most of which have become equipped with PCs and related peripheral devices such as printers. The size of these appliances has been shrinking and they have turned into integral components of desks in both offices and homes. Mobile phones are also small in size, so that it turns out that the ICT revolution has been the widest communications revolution with the least spatial requirements in both public and private domains.

The major spatial change brought about by ICTs is not in a reorganization of urban space but rather in the facilitation by these technologies of more efficient and sophisticated uses of any given urban spatial structure and organization as far as corporeal mobility is concerned. Thus, virtual mobility in cities complements corporeal mobility in various ways. Prior to a trip to an unknown location within a metropolitan area the trip can be prepared using Internet websites. The trip itself can be routed through GPS, whereas any scheduling problems emerging due to traffic jams can be negotiated with people in the trip destination through mobile phones. These ICT devices for personal mobility and their uses are indifferent to any existing urban spatial structures and organizational forms. The Internet and the mobile phone tend to kind of 'skip' over the immediate physical environment reaching remote locations of virtual destination, whereas GPS permits its bypassing whenever needed.

The social dimension

The possible lack of spatial reorganization in contemporary ICT-based cities does not imply also a lack of social changes at the individual level. It seems that a major change stemming from the massive use of ICT-based personal mobility devices is not spatial organization but 'human organization', in the sense that people now have to master several technologies in order to enjoy their spatial benefits, and even more strikingly, individuals have to get used to more flexible planning and organization of their time and schedules. Another required social change concerns a need to reformulate norms and manners for public behavior in light of growing virtual personal mobilities, notably regarding the use of mobile phones (see Kopomaa 2000, Ling 2004). Thus, the so-called domestication of ICTs (see Ling 2004) constitutes a double and continuous process of social and individual adaptations to the use of virtual personal mobility media in the public sphere, side-by-side with coping with new functions and options installed into new models of already-existing devices. Both processes of domestication may have implications for urban consumptions of ICTs. The first process which constitutes a social process of domestication, calls for reformulated norms of public behavior, and it may involve some potential limits on the usage of technologies in the public sphere, whereas the second process is rather a technological domestication process of coping with new technological options, and it may permit exactly the opposite, namely adding novel uses and applications of ICTs by users in the urban public scene. Thus, over time, some negotiation may emerge between widening and more sophisticated consumption of ICTs, on the one hand, and growing social concerns for their free and unrestricted use, on the other. Such a discourse has only just begun. Obvious examples are the frequent requests to avoid the ringing of mobile phones during concerts. It may take some time before such a discourse between social and technological domestication processes will reach saturation in the future, notably regarding the more problematic public sphere.

Conclusion

We have noticed several trends regarding ICTs and urban spatial reorganization. First, human beings are not ready, and maybe will never be ready, for a potential dismantling of cities when ICT-based media for personal mobility are ubiquitously adopted. Second, ICTs for virtual mobilities do not present any extensive spatial requirements for their very functioning. Third, media for virtual mobilities are indifferent to existing urban structures and urban spatial organization of any kind. Fourth, the media for corporeal mobility in cities have not changed in recent years side by side with the introduction and wide adoption of novel mobile media for virtual mobility. Fifth, city dwellers and visitors, equipped with mobile phones and GPS may move about any urban space in smarter and more efficient ways without regard to its spatial structure. Sixth, urbanites in the information age are in a process of change of habits as far as time use and technology use are concerned.

The kind of stagnation in the spatial reorganization of cities due to the development of ICTs for personal virtual mobility may potentially be compared to the life cycles of physical landscapes undergoing erosion. The classical, though heavily criticized along the years, model by Davis (1989, originally 1899) assumed three phases in the geomorphic cycles of physical landscapes: young, mature and old. The question is if we can observe a similar mechanism of change in cities as well, so that the current situation of no induced processes of spatial expansion will be interpreted as an old age phase in city development, maybe even signaling their future death. It seems, though, that this is not the case due to the very nature of cities as satiating a basic human need for places for aggregation. Cities have undergone spatial growth through mechanisms introduced into them by mobility technologies, but these mechanisms are not necessarily a must for their very existence which came into being millennia ago. By the same token, cities have turned out to meet a basic social and economic need for proximity and aggregation resisting any process which could have potentially dismantled them. Urban future renovation and change processes may, thus, evolve independently of any requirements for space by current technologies and independently of the spatial facilitation provided by them. As Mok, Wellman and Carrasco (2010: 2781) recently noted: "The combination of face-to-face, phone and e-mail communication means that the role of cities as interaction maximisers remains, in modified form. Cities continue to foster face-to-face contact and much contact is local. There is no global village. Rather, there is glocalisation, with extensive local contact joined by amplified long-distance connectivity. The city is no longer the boundary – if it ever was: it is the hub".

Chapter 10

Terminals

The previous chapter examined cities in general in light of new technologies for personal mobilities, and this chapter will rather turn to a focus on specific mobility facilities within cities, namely spatial terminals for mobility. These have maintained their importance even at times of growing mobilities (Knowles 2006).

We will explore first central railway and central bus stations, followed by a wider elaboration on airport terminals notably international ones. Central train and bus stations are terminals for terrestrial mobility modes only, whereas airports serve mainly as flight terminals, but they are also complemented by terrestrial mobility media (e.g. train, bus and taxi stations). Both railway stations and airports also normally include terminals for virtual personal mobility in form of Internet stands. Thus, the equivalent terminals for 'passengers' (users) of virtual Internet traffic are neither Internet exchanges nor Internet hotels, since the functions of these facilities are similar to those of control towers for the synchronization and coordination of air traffic as well as those of control centers for train traffic. The equivalent terminals for Internet traffic are rather the millions of users of the Internet themselves in all the varieties of Internet access options: in public spaces devoted for such traffic, such as Internet cafés or Internet stands; at home or at work using sedentary computers; and on the move using laptops or smartphones (on homes and offices as terminals see Kellerman 2006a, and on mobile users see Chapter 7 above). All of these computers constitute exclusive terminals for personal virtual traffic of all kinds (audio, visual, and textual), and this applies also to each user of public Internet access points, such as cafés. Thus, terminals for corporeal terrestrial and aerial mobilities are typified by a wide presence of vehicles, passengers and staff, whereas the equivalent terminals for Internet use can be either fixed or mobile, and they would normally consist of one 'passenger' only. The spatially fixed terminals for terrestrial and aerial mobilities will be assessed in the following sections from various perspectives of daily mobile passengers, their direct environment, and their wider spatial implications.

Central railway stations

Central railway stations emerged in the second half of the nineteenth century in what Eichards and MacKenzie (1986: 2) called "The Railway Age". Railway stations were originally located in sites which at the time of station construction were bordering city centers and later on, with urban expansion, have become integral parts of city centers themselves (Bán 2007). As numerous scholars have noted, these locations turned the stations into functioning simultaneously as transportation nodes and urban places (see e.g. Bertolini 1996, Bán 2007, Reusser *et al.* 2008, Trip 2008). A *node* in this regard was defined as: "a point of access to trains and, increasingly, to other transportation networks", whereas a *place* was defined concerning central railway stations as: "a specific section of the city with a concentration of infrastructure but also with a diversified collection of buildings and open spaces" (Bertolini and Spit 1998: 9), thus making central railway stations into "one of the most complex social areas" (Bán 2007: 289). The globalization of business which focused frequently in downtowns of major cities, side-by-side with the introduction of high-speed trains, added to the magnitude of the double nature of central railway stations as nodes and places (Trip 2008, Bertolini 1996).

Contemporary central railway stations, mainly in Europe, usually serve as termini for all levels of train lines: local, regional, domestic and international, frequently divided among several surface levels. They further serve as destinations and origins for many bus lines as well as taxis. Since their original development they have attracted, as places, cafés, hotels and shops on one side of the station, as well as freight yards, industry, and cheap housing on another one (Bertolini 1996). These land-uses have been complemented more recently by business and entertainment centers, again potentially divided between several sides around railway stations. Conflicts might emerge among these facilities from the perspectives of both users and land-uses. For instance, homeless people who share the same space with businessmen, and commerce and transportation infrastructure competing on the same piece of land (Bertolini 1996, Reusser *et al.* 2008). Bertolini (1999) claimed that in the long run the double functioning of the stations, as nodes and places, may reach a balance, and Reusser *et al.* claimed that such balances are rather dynamic and made possible by passengers who may use stations and their environs alike for various purposes.

Railway stations themselves were argued to present "a microcosm of the urban life" (Bissell 2009: 174), even more than airports, since stations are integrated into urban space with city residents and visitors walking through them. Bissell (2009) further argued that passengers constitute an integral part of stations, and their activities and behavior do not amount merely to movements across the surface of stations. The behavior of passengers in railway stations and their integration into them depend to a large degree on the luggage they carry for their travel, as this might limit their freedom of movement and shopping.

In a later section of this chapter we will discuss international airport terminals as terminals for aerial transportation, but it is important to point now to the major differences between central railway stations and airport terminals as transportation terminals (see also Bertolini and Spit 1998). Railway stations are not only located within city centers as compared to airports which are located at growing distances out of town, but they also constitute an integral part of city life, not disconnected from city centers by distance and by various forms of passenger controls. Railway stations are, therefore, more user-friendly than airports. At airports, hotels are in many cases located outside the terminal areas which is almost always the case for hotels in city business centers, so that both land-uses require travel to them from airport terminals. Residential areas are normally completely absent from airports due to the noise which accompany plane traffic and the transient nature of airports. Airports were termed by Augé (2000) as non-places as far as their identity and the social relations lacking in them are concerned, as compared to the role of stations as places which are located within cities, thus constituting integral components of city centers even if lacking residential populations around them (see also the discussion in the sub-section on airport socialities).

Central railway stations are usually used by train passengers on a daily basis, or generally more frequently than airports are used by most of their passengers. Thus, retail businesses around stations tend to offer more frequently-purchased merchandise, such as food and clothing, rather than the emphases on duty-free stores in airport terminals on items, such as luxury clothing, liquors and electronics. Also, the number of stores around central railway stations might be larger than the number of stores normally found in international air terminals.

Even glamorous central trains stations are normally less striking and more incorporated within the cities which they serve, as compared to the efforts to design tall and impressive airport terminals competing with each other internationally. Many of the central railway stations, notably in Europe, present impressive architectural monuments as perceived in the second half of the nineteenth century or the early twentieth one (Richards and MacKenzie 1986). Railway stations and trains carry with them some flavor of nostalgia, maybe excluding high-speed trains, whereas air traffic aims at the opposite, namely presenting an image of most contemporary technologies. Thus, the relationships between the mobile and the sedentary are stronger in central railway stations in the sense of business centers located in city centers as creators of human mobility in and around them, followed by mobile people as creators of sedentary mobility-oriented businesses in and around the stations. Similarly to airports, trains and railway stations incorporate ICTs in their operations, in form of information provision to customers and ticket purchase through the Internet, as well as system control in the stations and on the trains.

Central bus stations

Interestingly enough, central bus stations have not been able to create such a deep imprint in city life of European countries as central railway stations have, both as terrestrial mobility terminals and as urban places. This may stem from the stronger role of trains in both domestic and international long distance public mobility, as compared to buses. In North America too, central bus stations do not play a major role in urban landscapes. This may stem from the preference for air and car mobilities there. Generally, trains emerged earlier than buses, requiring much more space than buses and were perceived at the time as progress symbols, so that more efforts were invested in the design and construction of central railway stations as compared to those for buses. In South America, though, central bus stations may frequently play a more dominant role in urban life, such as the case of the Brazilian *Rodoviarias.*

The literature devoted to the role of central bus stations is almost nonexistent. An exception to this lacuna is the new central bus station in Tel Aviv which attracted some attention, given the preference for bus transportation for both local and inter-city travel in Israel, and since this station might be considered the largest central bus station in the world (Amdur and Epstein-Pliouchtch 2009). The station, which was opened in 1993, stretches over 250,000 square meters on eight floors and it was planned as 'a city under a roof'. However, despite of its daily 70,000 users many of its stores are closed, given the location of the station in a poor neighborhood of labor immigrants. However, its users perceive of it as a place, and partially at least disconnected with the city, serving more as the center of the country. Thus, a bus station may be perceived as a place connected to regional and national spaces rather the than the adjacent local urban ones, reflecting simultaneously the local context of the station and its service areas.

International airport terminals

Airports have been considered as "an ideal place to consider the ways in which geographies of human mobility have developed" (Cresswell 2006a: 220). Thus, airports in general, and international ones in particular, have received the widest attention in their role as mobility terminals in recent mobility literature, and we will, thus, devote to them extensive attention here as well (Kellerman 2008). Flights and airports were considered as replacing railways and railway stations in their current centrality for travel and tourism (Richards and MacKenzie 1986). As such, airports constitute most significant terminals for routine business travel as a 'daily' form of mobility.

What are international airports all about? Simply answering, they constitute break-of-bulk sites between terrestrial and avionic modes of transportation, facilitating the boarding and disembarking of passengers and the loading and unloading of commodities, going to or coming in from other countries (see Adey 2007). It so turns out, however, that international airports, and notably their passenger terminals, have been considered to constitute much more than this rather functional definition of their service. Thus, airports have been variously viewed as constituting or as signifying and meaning "national frontier on the outskirts of a major city in the middle of a country" (Pascoe 2001: 34, see also Salter 2004); as *Aviopoli,* as *Aerotropoli,* or simply as cities (Aaltola 2005, Gottdiener 2001, Kasarda and Lindsay 2011, Lloyd 2002, Fuller and Harley 2005); as thresholds and transition spaces (Gottdiener 2001); as a pedagogical place educating passengers to hierarchical order (Aaltola 2005); as a commercial interchange without human meetings (Crang 2002); as non-places (Augé 2000); as "code/space – space produced through code" (Dodge and Kitchin 2004: 204); as "iconic space for discussions of modernity and postmodernity" (Cresswell 2006a: 220); as omnitopia (Wood 2003), as expressing air-mindedness (Adey 2006b), as "sites of surveillance" (Lyon 2003: 14), as "symbols of mobility" (Adey, 2004b), as "mythical place of promise" (Fuller and Harley 2005: 105), as "'vessels of conception' for the societies passing through them" (Pascoe 2001: 10), and, finally, as "no-(wo)man's land" (Braidotti 1994: 18).

As we will see, airports seem to simultaneously constitute all of the attributes mentioned so far, attributes which focus mainly on perspectives relating to the largest group of airport users, the passengers. It is important to realize, however, that airports can be viewed from several other perspectives as well: first from the perspective of the suppliers and supply in general, namely airlines and flights. Second, airports may be looked upon from the perspective of regulators or governments and airport management bodies. This perspective cannot be completely separated from that of passengers, since regulators have established growing authoritative frameworks for airport management and operations dominating all other airport dimensions. Adey's (2006b) notion of *air-mindedness* relates primarily, but not only, to governments. Third, airports constitute an expansive land-use located at the outskirts of major cities, and airports which serve as hubs also facilitate additional land-uses of plane maintenance, food caterers, storage areas, etc. in and around them (see Cidell 2006). As such, airports may be interpreted as kind of theme parks mixing various retailing facilities, albeit simultaneously serving also as potential gates for detention in their role as border-crossing posts. As such, they constitute a sort of extraterrestrial points, located in the heart of countries (Lloyd 2002).

International airports constitute, above all, spaces of highly explicit expressions of authority by several bodies, expressed through the terminal environment, its operations, and through authority-generated flows. Thus, these authorities determine, directly or indirectly, both the fixed and the mobile within airports at large and within airport terminals in particular. These powerful expressions of authority stem not only from the function of airports as border stations, but also from the big airplanes served by airports, requiring authoritative management, both in the air and on ground. We will further argue that airports constitute complex social places involving special attitudes and dialectic perceptions and modes of behavior by passengers.

A clear differentiation has to be made between airports which are more domestic in nature, and those which are more international. A busy airport does not necessarily imply its being mainly an international one. International airports differ from domestic ones in a large variety of authoritative controls and services which exist only in international airports: passport and customs controls, duty free stores and currency exchange services, to mention just the most striking ones.

Airports were probably first established in the US in the 1920s, and the first terminals in them were built in the 1930s (Gottdiener 2001). "Flight connected an American national identity and a dream of freedom and unfettered movement" (Adey 2006: 346). Right from their beginning airports were space-rich because of their runways, and the space needed for aircraft parking and taxing. Contemporary airports operate wide-body big planes, which need longer runways and more ample gate-space, coupled with a tremendous growth in the volume of air passengers. They have required, therefore, much more territory, and have thus been frequently located more remotely from the cities which they serve, sometimes even built on artificial islands, such as in Hong Kong and Osaka. Airports, have, thus, been termed 'terraformers' (Fuller and Harley 2005: 102).

Despite the relative glamour attributed to international flights and provided by enhanced terminal design, duty-free stores, and onboard services, international passengers constitute a minority in the world total of air passengers. IATA (International Air Transport Association) (2006) reported for 2004 a total of 1,888 million domestic and international air passengers, whereas UNWTO (UN World Tourism Organization) (2006) reported for the same year some 330 million international tourists arriving by air, which seemingly amount to at least 660 million flights or some 35 percent of total air passengers for 2004. However, arrivals may represent several flights per tourist, so that sometimes domestic flights are continued with international ones or *vice versa*, and sometimes all the tourist's flights are international. It might, thus, be safe to assume that international passengers may amount to a maximum of 40 percent of air passengers. The need for domestic connecting flights to international ones raises the indirect volume of internationally induced air travel, maybe reaching one half of total passengers. It is important to realize that even in a globalized world of intensive corporeal and virtual connections, still more people fly domestically than internationally. As we noticed already in Chapter 8, some 25 percent of global international tourists are business ones, but we can't extrapolate from this datum on the ratio of business visitors among air international air passengers, since the estimate of 25 percent business travelers relates to business visitors in general, including arrivals by all corporeal means via air, ground and sea.

Large-scale and frequently hub international airports normally comprise several functions beyond runways, terminal(s) and the numerous facilities which the latter house. These additional airport functions include mainly aircraft maintenance areas, car parking lots, catering supplies, large-scale storage areas, and more. As such, airports constitute not only a major land-use but also a major employer. It was estimated that London Heathrow Airport (UK) employed some 55,000 people (Amin and Thrift 2002), and Chicago O'Hare Airport (US) employed 50,000 people in an area of 8,000 acres (Gottdiener 2001). Furthermore, airports constitute major business beyond the basic fees collected from airlines and passengers, given all their related services, such as parking, storage, shopping and eating facilities, etc. (Francis *et al*. 2003, 2004).

Airport geographies

Through the lenses of mobility, airports may be viewed as having their own, and rather multifaceted, geography. This geography consists of four major facets: first is the *nature and the 'identity'* of this special and space-consuming facility as an authority-driven and operated one; second is the *fixed environment*, namely the spatial organization of airports, and even more so of their passenger terminals, and the powers which both shape and operate them; third is the *mobile*, referring to terrestrial movements of passengers within the airport (as well as their avionic movements, relating basically to the number of domestic and international passengers served by each airport); and fourth is the *social*, attempting to understand human perceptions and behavior in airports.

The geography of airports, by the very nature of airports, cannot focus merely on airports, possibly ignoring the kind of 'nesting' and 'anchoring' of airports within the cities and countries which they serve (see e.g. Cidell 2006). Volumes of passengers, their origins and destinations, and their purposes of visits, all depend on the city served by an airport: the cities and countries with which it is connected by air; whether the connections have a dominant tourism sector (business, leisure, family visits, etc.); and the volume of passengers traveling on specific lines. Furthermore, if airports are assumed to constitute aviopoli, then it is important to learn how they resemble and in which ways do they differ from 'normative' residential cities. Airports also reflect the countries which they serve. This may be evident in some of the foods served in airport eateries, as well as in some of the merchandise sold in duty-free stores. However, as we will see later, country identities are most visible in international airports through the use of domestic-national languages in airport signs.

Our discussion of the four facets of airport geography will begin with an elaboration on the various airport authorities, outlining their roles and interrelationships, and exposing them as the predominant characteristic of international airports (Figure 10.1). This discussion of authority will be followed by elaborations on terminal environments and their operations, on passenger flows, and finally on passenger socialities, accentuating the roles of authorities in their shaping. These dimensions seem to jointly express the fixed and the mobile in airport terminals. The physical terminal environment represents the fixed element of airport terrestrial and avionic mobilities, and their 'operations' implies their functionality. The operated terminal environment produces terrestrial mobilities about the terminal, namely passenger flows. The physical environment, its operations and the resulting flows, are believed to enhance, facilitate or deter passenger socialities. The following discussions will focus on international airport terminals only, rather than on the whole of international airport complexes, because we assume that terminals are the structures through which outgoing and incoming passengers are served, and in which most of the services offered to travelers are located.

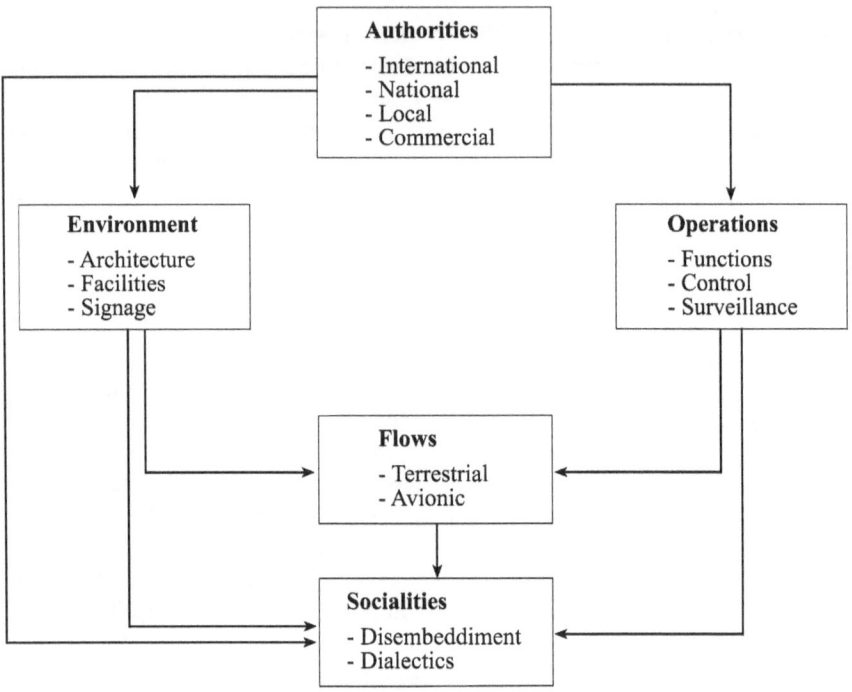

Figure 10.1 Facets of international airports

Airport authorities

International airport terminals may be considered as constituting the most authoritarian facility designed for the use of free civilians, typified, as we will see, by a wide base, amount, domain, and scope of authority powers. As such, airports are both *authoritarian*, issuing commands to be obeyed automatically, and *authoritative*, issuing well-explained rules and regulations (see Wolf 2001). Probably only prisons and military bases, meant for confined prisoners and mobilized military forces respectively, are as authoritarian as international airports. "Antiterrorist measures turned the airport into an electronically controlled environment rivaled only by the maximum security prison" (Gordon 2004: 238).

The glamorous design of terminals, the entertainment and shopping arcades which they house, and the vacation atmosphere which they seemingly inspire, all but conceal the simple fact that several levels of strong authorities, international, national, local, and commercial coerce a wide base of both positive and punitive powers on passengers, as well as on workers at international airport terminals, bringing about compulsory behavior in a large number of people. The punitive powers are instant and on-the-spot, covering a wide scope of possible punishments: short- or long-term boarding or disembarking prohibition; refusing to sell in duty-free stores; or confiscating products in customs. The controlling power extends to all the major dimensions of terminals or the airport domain: their fixed environment; the operations of preparatory functions for air travel; flows of passengers within terminals and the movements of planes around them; and finally also to passenger socialities. For Aaltula (2005: 261), airport authoritarianism carries also a pedagogical mission for modern social life at large: "Airports are places where authority is recognized and instructions for making 'proper' judgements and acknowledgements are given. It can be hypothesized that airports teach people the central rituals of acknowledgements that are needed to navigate in the Byzantine structures of the modern hierarchical order."

Salter (2003, 2004, 2007) pointed to two conflicting governmental motivations in airport functionality and authority: maximum security and free mobility. Another, and rather secondary and local conflict, may also typify airport managements: speedy flows of passengers in terminals versus money spending by them in duty-free and other commercial facilities (Adey 2007, Francis *et al*. 2003, 2004). Whereas the second conflict may be balanced almost completely by passengers using watches for their time management, the first one is almost solely dependent on national and airport authorities, though here too passengers are supposed to police themselves, through their obedience to security regulations, avoiding power exercising (Salter 2004).

Authority relations at large emerge when a common ground between rulers and followers exists (Friedman 1990). One must further assume the prevailing of a *dependence thesis*, namely that followers have reasons to obey rules and regulations even before authority exerts its power (Raz 1990). For air passengers, mobility is obviously the most important desire, but they would like to fly with maximum safety and security. National governments are, foremost, concerned with wider national security coupled with specific security and safety concerns for each flight. Authority in airports constitutes, therefore, a complex system which governs both security and safety (or mobility), and simultaneously mediates between them. The tightening of security measures in recent years has been based on a balancing between mobility and security, but even more important, these procedures have been viewed as such by passengers, who have continued to fly despite the heavier security measures. Alternatively, imbalances between the two motivations of mobility and security could have resulted in less passengers (by individual decisions) and/or in public objections through constitutional and media involvement (bringing about collective decisions). Another tool for balancing security and mobility has been the growing commercialization of airports, offering shopping and entertainment compensations to passengers 'in return' to stricter security measures imposed on them.

As we mentioned earlier, international airport terminals are managed, operated, and controlled by various levels of authority: international, national, local, and commercial, and as such they present diffused and shared power and authority without a 'real' source of power (Salter 2007). We will briefly outline each of these authorities. At the international level, the extensive activity of foreign airlines is imminent for international airports, coupled with the constant presence of foreign territories in the form of the interiors of foreign airplanes. These foreign territories of airplanes are mobile, and their very location and extent within airports are under constant change, but they carry bold national symbols and/or colors, and they are open to large numbers of selected citizens possessing boarding passes. Thus, foreign airplanes in airports present a much stronger routine foreign spatial presence as compared to concentrations of foreign embassies on selected streets in capital cities, which administer restricted access to the general public of the hosting countries. Sometimes an even stronger foreign, and rather fixed, presence may exist in airport passenger terminals. For instance, the US maintains passport and customs control services for flights to US destinations within the Toronto, Vancouver, and Ottawa international airport terminals, thus permitting the channeling of incoming flights from Canada into domestic terminals/gates at their US destinations. This procedure further reduces the international dimensions of American airports which are characterized by heavy domestic flight activity.

National authorities have always been at the core of the airport business, since airports serve as international border-crossing passes inside the country (Pascoe 2001, Lloyd 2002), and those located in capital or leading cities provide access to centers of political power (Aaltola 2005). Passport controls and surveillance by national authorities are, therefore, highly required. The very extensive presence of foreign planes and airlines calls for special attention and vigilance, and the very service of an international airport for routine international ties with other countries may constitute crossroads, and has even been considered as a substitute for military occupations (Aaltola 2005). In addition to security and control, the flight business requires the highest levels of safety, coordinated by both aviation agencies and ground emergency forces. Furthermore, and at the financial sphere, airport terminal shopping arcades contain two duty and customs zones, those charged with duties and those free of, and this separation calls for control by additional national agencies. All these powers of national authorities in airports have gradually increased since the 1970s because of plane hijacking and airport terrorist attacks, and they have become much tighter following the 11 September 2001 attack.

Local airport authorities are normally in charge of the smooth operation and maintenance of airport terminals, and since airport terminals are active almost all day and night, seven days a week, serving thousands of passengers simultaneously, much authority, service and guarding are necessary. Local airport authorities may share several of their responsibilities with national authorities.

Commercial international airlines are too involved in passenger services implying strict authoritarian care. First, because they operate a rather costly business, and thus for most airlines single tickets are normally charged hundreds of Euros, or their equivalents. Mistakes in check-in and/or boarding can be costly for both airlines and passengers. Furthermore, airlines have to adhere to strict safety measures, as far as size and weight of luggage is concerned. They also have to ensure that passengers hold passports and visas fitting their destinations.

Environment

Following the discussion of authorities as the dominating element in international airport terminals, we move now to the four sociospatial dimensions of terminals (environment; operations; flows; and socialities). First of these is the physical airport terminal environment, or its fixed dimension, which in itself may be viewed as consisting of three major dimensions: architecture or design; facilities; and signage, all of which represent or involve some kind of authority which determines location and functioning within an authoritarian environment. From a functional-architectural perspective contemporary international airport terminals combine under one roof two major classes of facilities: flight related ground services of checking-in, boarding, disembarking and checking-out, as one class of services, and a wide variety of auxiliary urban services, mainly eateries, shopping, entertainment, personal care, meditation, accommodations, etc., as another class of services. In this aviopolis, architects and decision-makers who direct architects determine the strict location of any facility, its size, and its access, as well as its times of operation, in ways that no urban planner or local municipality can or even want to achieve.

This roofed, strictly organized, and rather comprehensive urban space has come into being particularly since the 1970s with the introduction of wide-body planes, permitting much larger numbers of passengers per flight, and requiring much more space for both the planes themselves and passenger services of all kinds. The technological change in plane size was coupled with economic growth in developed countries thus bringing about more air travel. Simultaneously, information technology was heavily disseminated in the travel business, permitting the operations of the physically expanding terminals and the flight business in general. Hence, the accreted terminal was replaced by large scale terminals, connected with gate areas via pedestrian or rail concourses (see Gottdiener 2001). A somehow opposite contemporary trend has been the designation of separate regional, low-cost airports located mostly in European cities (e.g. Bremen in Germany and Marseilles in France). Such low-cost airports are meant to serve low-cost airlines, and are thus typified by more modest infrastructures and services, bringing about lower costs for airlines (see Francis *et al.* 2003, 2004).

Three principles have been proposed for the architectural design of the contemporary concourse-based terminals from passengers' perspectives (for a historical review see Pearman 2004). First, for travelers' minds, is the *memory principle*: "Airport architecture assists remembering and memorizing by containing an overall consistent form together with striking parts. In this regard, the airport is a vehicle of memory and understanding" (Aaltola 2005: 271). Second, for passengers' feelings, is the *transition principle*: "Entering the space triggers new feelings of self, new identities that are set off by stimulators...the airport is the definitive transition space. Trips are nothing if they are not existential" (Gottdiener 2001: 10). Third, for travelers' experience, is the *newness principle*: "The airport is a new kind of space, a new kind of experience, and a remodeling of some very old aspects of the built environment...many passengers actually spend more time within airports per trip than they do at the bank or even the supermarket" (Gottdiener 2001: 61). These passenger-oriented principles are coupled with authority-oriented principles, assuring strict governance of the terminal, free flows of armed forces, and fast evacuation options.

Though the urban facilities and services offered in terminals are varied, similarly to those available in cities, they tend to be more spatially concentrated by the very physical structure of terminal complexes as compared to urban settings. Another major difference between cities and international airport terminals is the lack of permanent residential population in the latter, other than the homeless. In several airports, passenger and crew hotels are physically annexed to the terminals themselves (e.g. in Frankfurt and Prague). The film *The Terminal* (Steven Spielberg 2004) interpreted a true case of an Iranian refugee who lived for eleven years in Charles de Gaulle Airport (Lloyd 2002). Focusing on the new concourse-based terminals, it tells the story of a passenger coming from an East European country which ceased to exist upon his landing at JFK Airport. He cannot be admitted to the US nor can he return home. The film presents the airport terminal as a compelled home and its employees and passengers as potential neighbors exposing rather grotesque scenes of a variety of airport authorities and scenarios.

A significant physical element in airports is signage, "often neglected in commentaries on airports" (Gottdiener 2001: 75). Like in city landscapes, airport signage is divided into two classes: commercial (by airlines, stores, and general advertising), and directional. Directional road signs in cities consist mostly of internationally recognized symbols or icons, occasionally also with minimal texts. In airport signage the international logo system is almost always coupled with texts, despite the rationale of the logo system to overcome problems of language knowledge by foreign visitors. Whereas road signs in cities accentuate the do's and don'ts, airport signs are much more directional, pointing passengers to specific facilities and locations, so that they will "be able to negotiate the space of the airport terminal with ease and rapidity" (Gottdiener 2001: 75). However, airport signs are not just directional in nature, because they also dictate the spatial flow of time-constrained passengers. If information implies power (Castells 1998), then this power is used to ensure the movements of passengers in accordance with the design and directives of airport authorities, and not just in order to provide for the way-finding welfare of passengers.

Conflicting assessments have been made regarding the potential service of airport signage for a possible creation of sense of place. For Gottdiener (2001: 75) they "contribute significantly to the creation of a sense of place in otherwise transient spaces", whereas for Cresswell (2006a: 244) "signs are a central part of what the anthropologist Marc Augé [2000] calls non-place. Non-places, often spaces of transit, refer to other places without taking you there." Some of this difference of opinion may relate to the question of language used for signs. For Gottdiener, who wrote on airports from the perspective of the American scene, it was obvious that US airports carry signs in English only, thus possibly creating a sense of place for American passengers similarly to signs on city streets. For Cresswell, writing from a European multilingual perspective, languages on signs is a more complex issue. Focusing on Amsterdam's Schiphol Airport, in which signs are in English only (ignoring Dutch) he concluded that "Schiphol may be a node in a global space of flows, but it is still uniquely Schiphol – still a place" (2006a: 257).

Signs may further be interpreted as 'cultural artefacts' (see Lash and Urry 1994: 3), since 'sign language' in international airports raises the question of balance between the domestic and the global. The use of domestic languages in airport signs provides a sense of place for residents of the host country, notably as compared to airports they are about to visit in their destination countries, or those they have just returned from. Domestic languages further provide for some national symbolism and national atmosphere in a facility, the international airport terminal, which is similar in its functionalities to airport terminals in other countries. On the other hand, the terminal provides services to passengers who mostly or mainly come from foreign countries and do not necessarily master domestic languages. Hence, "the lingua franca of the international traveler is not Esperanto, but English – the language used at airports and railway stations worldwide" (Mijksenaar 2003). Furthermore, English was the formal language of close to 22 percent of world tourists in 2003, with 8.9 percent of world tourists coming from the UK (second to Germany (10.8 percent) in the number and share of outgoing tourists), and followed by the US (8.1 percent) (Table 10.1).

Table 10.1 International outgoing tourists from English speaking countries, 2003

Country	Number of tourists (in millions)	% of world total
Australia	3,388	0.5
Canada[1]	17,739	2.6
Malta	0.174	0.03
New Zealand	1,374	0.2
Philippines	1,803	0.3
Singapore	4,221	0.6
South Africa (2002)	3,794	0.6
UK	61,424	8.9
US	56, 175	8.1
Total	**150, 092**	**21.83**
World total	**689.7**	

Note: [1] Including Quebec.

Sources: Table: Kellerman 2008. Data: WTO 2005.

How do airports balance between the domestic and the global in informative and symbolic signage? Examining several leading airports in non-English speaking countries reveals three patterns: a totally global pattern ignoring the domestic; a pattern treating equally the global and the domestic; and a third, giving preference to the domestic over the global. An example, and maybe the only existing case, for the first, totally international, option, is Amsterdam's Schiphol Airport, which we mentioned above and which carries signs in English only, with black letters on a yellow background. Examples of the second, more balanced option between the domestic and the international, are Frankfurt and Hong Kong, which portray blue signs with white letters, with German and Chinese, respectively, on top and English underneath, all languages being in equally sized letters. Examples of the third option, leaning more to the domestic, are Charles de Gaulle in Paris and Narita in Tokyo. In the first, rather extreme, example, signs are in yellow, with French on top using black, and thus, striking, letters, and with English underneath using white letters. Thus, the initial visual perception is of French signs. In Narita Airport, signs are in grey and lettering is in white with the Japanese text on top in slightly larger letters than the English texts underneath.

Operations

In this section we refer to the operationalization of the fixed terminal environment described in the previous one. The operations of an airport consists of the various functions performed in it, notably for passengers, the controlling of these and other functions, for instance take-off and landing, and surveillance against potentially illegal passengers and terrorists. From the perspective of passengers, they move from one service station, or check point, to another, on their ways into or out of countries, meeting each time another authority, so that the airport seems to be divided among these authorities, at least in terms of the service chain. Another division of terminal areas or services is the strict physical one between incoming and outgoing passengers. However, this division has been somewhat weakened with the diffusion of mobile phones permitting instant and continuous audio contacts between incoming and outgoing passengers. Major, and constantly growing, is the security checking and surveillance of passengers and their luggage, employing various security technologies and procedures (see e.g. Adey 2004a, 2007, Lyon 2003, Salter 2004, 2007), as well as the architectural adjustments of terminals required due to security measures imposed mainly following the terrorist attack of 11 September 2001(Petroski 2004). These measures also involve service quality concerns stemming from stricter security checks and surveillance (Gkritza *et al.* 2006), notably since airports are constantly in states of safety and security emergencies (Fuller and Harley 2005: 46). Interestingly enough, airports contain, on the one hand, the largest concentration of duty-free stores, serving outgoing passengers, but on the other hand, it is the only place where residents and visitors as incoming passengers have to pay customs and taxes, whereas normally this is the responsibility of sellers.

On the way out of a country several authorities check the passengers, and these checkpoints are spread throughout the terminal space. At the check-in counter, airline agents serving the passengers perform as both domestic and foreign commercial authorities. These agents check not only commercial documentation in the form of tickets, but also passports and visas, and they may further ask security questions. Passports play a special role for passenger identification, tracking and regulation throughout the airport (Salter 2003, 2004, 2007). In the next station, passengers meet local airport and/or national security forces, electronically checking their hand luggage and body. This is followed by national authorities checking passports and visas. These three stations involve some tension in passengers (Iyer 2001, Salter 2003). Passengers who choose to purchase at duty-free stores, have to present boarding passes to the cashier, who is an employee of a commercial company. Finally, once again, boarding passes have to be presented to airline agents on boarding the plane.

On the way into a country, the chain of authorities is shorter and involves only two national ones: passport and customs controls. These two stations again involve some passenger apprehension and attempts may be made, through their design, to relieve these tensions (Salter 2003). In the US, agents in these stations have more legal power as compared to authorities within the country (Salter 2003, 2004). Both incoming and outgoing passengers are asked questions, in person and in writing, viewed by some (Salter 2007) as 'confessions', and as 'contractual declarations' by others (Fuller and Harley 2005: 44). The only terminal functions/spaces which may be used by passengers without explicit authorization are the public lounges, eateries, entertainment facilities, meditation rooms, and toilets, which have been considered as areas of immobility (Adey 2007).

The contemporary flight business, based to a large degree on wide-body planes and huge numbers of passengers, could not have functioned without wide and deep dissemination of information technology, stretching from the obtaining of information by passengers via the Internet, through reservations made by them or by travel agents, to airport functioning (Kellerman 2006a). Dodge and Kitchin (2004: 197) termed airports "'code/spaces' – spaces in which the materiality of air travel is produced through information and communications technologies (ICTs) and software systems," and Cresswell (2006a: 238) noted that "in the airport, the construction of material space and the programming of software have become inseparable" (see also Lyon 2003). Side-by-side with the functional crucial value of information technology, it provides also immense control and management power to airport operating agencies. Whether it is the controlling of plane ground and air movements, or the controlling of passenger services at check-in or elsewhere, or be it the work procedures of employees, all are centrally and electronically controlled by computer and telecommunications systems. As such, airports, probably constitute the most IT-controlled public service facility.

Information technology constitutes also a major component in the surveillance systems in international airports (Dodge and Kitchin 2004, Adey 2004b). Passengers are under surveillance almost everywhere in the airport, as well as on board, controlled by overt and covert means and equipment. In this, once again, airports might be considered the most guarded public facilities, presenting authoritative power largely hidden to innocent passengers.

Flows

The operation of the fixed airport environment is aimed at the production of flows. By their very nature airports imply avionic flows of planes in the air, as well as on the ground, through taxiing, take-off and landing. Side-by-side with these vehicle movements airports imply the terrestrial flows of passengers, whether through their simple walking or through the use of beltways, elevators, escalators, or inter-terminal trains and buses. These flows are heavily determined and manipulated by authorities, using the national passport and the international boarding pass for flow regulation (Fuller and Harley 2005: 43), and guiding passengers through airport signs. Furthermore, authorities have an obligation to airlines to provide speedy processing of passengers, but, on the other hand, they are interested in passengers' spending, which has become an important source of income especially at times when low-cost airlines pay less for airport services (Francis *et al.* 2003, 2004, Adey 2007).

Compared to people's movements about cities, airport passenger flows may be distinguished by four major aspects: directionality; controlling; separation and timing. By directionality we refer to the specific directions that have to be taken by outgoing passengers along their way to their flights or by incoming passengers proceeding towards the terminal exit. These directions, specified by the various stations/functions required by passengers before their boarding or exiting, and informed through terminal signs, provide for "regulated space in which mobilities are closely and carefully channeled" (Cresswell 2006a: 248). In cities too there is some directionality, but usually cities permit several routing and mobility alternatives using a number of roads to get to any specific destination, and the variety of local destinations of city dwellers and visitors is much wider than those of passengers aimed at reaching even numerous gates in airport terminals. It is again authority, as well as flight specifics, which determine airport passenger directionality.

Controlling is a second major aspect of human flows in airports. In cities, people are supposed to flow freely, whether as pedestrians or as drivers/passengers. When the police stop somebody on an urban street for identification it implies suspicion of some wrong-doing. In airports the opposite applies: all passengers are stopped at various stations for strict identification. It is rather that any 'freely' moving passengers who do not stop where required to do so immediately invite severe suspicion and possibly immediate detention.

A third major aspect of airport human flows is separation, and not on a group basis, e.g. by concourses leading to different flights, but rather on an individual basis, so that certain people may be stopped at certain stations while others are permitted to go ahead (Aaltola 2005). This is strikingly evident, for instance, at the customs control station while leaving the airport.

Human flows in terminals are further characterized by their strict timing, or their time contexts and framework, in that the flows of outgoing passengers are determined by flight schedules. While in city flows people can allow themselves to be late reaching some of their destinations and objectives, in preparations for flights in terminals this possibility is extremely restricted. Thus, passengers are frequently in a hurry when moving about terminals, either because of their late arrival at the terminal, or because one of the stations, notably check-in counters, are congested. At the other end of the flight process, on passengers' way out of the terminal, the differentiation of people by their timing is not through flight schedules but rather on a personal basis, similar to the time constraints of people moving about cities.

Socialities

Like other terminal dimensions airport passenger socialities may too be viewed as facilitated or as restricted by airport authorities, but socialities are also managed by the other terminal dimensions, namely environments, operations, and passenger flows. It is important to note from the outset that even before encountering any airport authority, the very entering of passengers into an airport terminal on their way to another country implies that they are being disembedded in the most literal physical sense. At the specific site of airport terminals they are neither at home, nor at their destination, and not even on the vessel, the plane, which is about to take them there. The status of travelers as transitional or de-territorializied (Gottdiener 2001: 32) is at its utmost in terminals, making them feel as if their nationalities were abolished (Cresswell 2006a: 222). Upon arrival, being again at an international airport, this status is softer and shorter, since the process of checking out of airports involves fewer stations and is, thus, normally shorter than that of checking in, since no plane boarding follows, but rather terrestrial movement to either their destination or home.

Airport terminals constitute fixed physical environments of disembeddiment from origins, from physical mobility vessels, as well as from destinations, but they still provide some resemblance of home through the shopping malls and food courts (Iyer 2001). On the other hand, terminals have turned into highly connected complexes through the ample provision of electronic communications media, permitting virtual connectivity via all possible media, in some way also in order to compensate passengers for their physical transitional displacement. Thus, airports once offered numerous public phone booths, and more recently terminals turned into the first locations for the installation of Wi-Fi 'hot spots' for wireless Internet connection for passengers while in airport terminals.

Several notions have been proposed over the years for the unique sociospatial terminal contextualities: *non-place*; *placelessness* and *omnitopia*. The rather transitional nature of airport terminals, despite the amenities which they offer to passengers, and their lack of home-based residential population, have made Augé (2000) call them non-places. "If a place can be defined as relational, historical and concerned with identity, then a space which cannot be defined as relational, or historical, or concerned with identity will be a non-place" (Augé 2000: 77-8). The idea of non-place reminds one of the much earlier notion of placelessness, originally proposed by Relph (1976). However, whereas Augé accentuated more the lack of *social* relations in his notion of non-places, Relph emphasized the *personal* lack of sense of place (Kellerman 2006a). Cresswell (2004: 45-6) noted in this regard that non-places do not carry "negative moral connotations" whereas the idea of placelessness does. Merriman (2004) criticized the notion of non-places, arguing that the production of environments of transience involves complex social relations among various airport operators and workers. Thus, "airports are a new kind of space that provides portals to the realms of both place *and* placelessness" (Gottdiener 2001: 61), and by the same token airport terminals may be viewed as constituting simultaneously both places *and* non-places (Adey 2006b).

One may view passengers as devoid of social relations: "In the middle of the cold beauty of this airport passengers have to face their terrible truth: they are alone, in the middle of the space of flows" (Castells 1996: 421). However, feelings of temporary placeness and sociality may emerge when flights are delayed, bringing about interactions among passengers who did not interact with each other before the delay, thus creating temporary communities (Gottdiener 2001: 30-1).

A third interpretation of terminal social spatiality was omnitopia, proposed by Wood (2003) as representing "the construction and performance of geographically distinct spaces as perceptually ubiquitous place" (p. 325). As such, terminals are supposed, by Wood, to create "generic environments", to invoke "continual movements" and to enact "atomized interactions" (p. 327). Feelings may differ, though, from passenger to passenger (see Salter 2007).

Passengers in international airport terminals on their way to another country are at a unique location in which they experience three major inner or outer forces: disembediment; authority; and expectation. These rather contradictory forces imply several dialectics. On the one hand, travelers are disembedded from their home environments and routines, and at the same time they have to function and move under strict authoritarian procedures throughout the terminal. These two contradictory forces are coupled with expectations for both the trip and its destination, and as Relph (1976: 87) noted on the perceived pleasure attached to the flight experience *per se*: "where someone goes is less important than the act and style of going." Furthermore, "airports are the gateway to the 'other', i.e., another staging area of action ... A trip, any trip, provides us with the freedom to escape our daily routine" (Gottdiener 2001: 11). These good feelings of pleasure and expectations of more pleasure to be experienced at the destination are coupled with anxiety at the airport, stemming from both its authoritarian nature and trip apprehensions, such as potential delays, losing of valuables, etc. (Aaltola 2005).

Airport terminals further imply a simultaneous mix of social separation and blending. On the one hand, passengers are separated by flight classes at various stations: check-in, lounges, boarding, and sometimes also at security and pass-control. On the other hand, international airports are great mixers of people of different cultures, nationalities, professions, etc., more strikingly than city streets (see Cresswell 2006a, Aaltola 2005). Passengers are supposed to keep social distance from each other, under normal short terminal stays, similarly to norms prevailing in urban public spaces. However, airport terminals may offer opportunities for social transactions, not only because of flight delays, but in formal conference rooms, as well as informally at the various lounges and eateries (see Gottdiener 2001).

Conclusion

Transportation terminals turn out to be a must for all forms of public transportation, whether terrestrial, aerial or maritime. It further turns out that their impact on urban life may be a double one in their constitution as both nodes and places. This has been argued for central railway stations, and it might be true, at least partially, also for central bus stations. Airport terminals are surely transportation nodes, and as it was shown, they might potentially be considered as places or non-places. From the perspective of commercial and business activity in international terminals they might, in many airports, be similar to shopping malls. Despite their location outside of cities, airport terminals, notably international ones, have a special importance for contemporary mobility, which has growingly become based also on aerial travel. Airports tend to function under strict controls, notably security control, thus serving as a special spatial entity of human activity.

One may discern several recent processes in airport terminals, notably international ones, worldwide. The most striking is the standardization of equipment and procedures, required by the large volumes of passengers and by growing airline acquisitions and code-sharing agreements. Procedure standardization has been made possible by extensive adoption of ICTs. Thus, luggage is treated equally worldwide at check-in stands and in luggage claim carrousels, boarding passes of different airlines contain similar items of information, and security screening equipment is similar as well. In addition, the growth of multinational eating and shopping chains adds too to the standardized looking of airports, coupled with the use of English in signs. Differences in terminal design and furniture may often offer the only specific remarkable characteristics of airports. International airport terminals seem, therefore, to constitute the only truly global standard public facility. Similar processes of standardization, though weaker ones, also characterize central railway stations, notably those which serve, mainly in Europe, international trains. Such standardization may apply to electricity lines for trains, signs, tickets, etc. This standardization is enhanced by the concentration of locomotive and car production by a small number of manufacturers.

Another contemporary process is the intensification of security or authoritarian processes in airports, because of growing terrorist attacks and threats, taking place side-by-side with the widening offerings of airport entertainment, through casinos, spas, more varied shopping facilities, etc. International airport terminals have become, thus, simultaneously the most guarded and most varied consumption complexes under one roof. In central train and bus stations security measures are not as extensive as in airports, and the entertainment and commercial facilities connected with airports are not necessarily located within stations, since stations are completely integrated into the urban commercial web surrounding them.

Chapter 11
Opportunities Through Daily Virtual Mobilities

The two previous chapters in this third part of the book focused on impacts of daily spatial mobilities on urban space *per se*, exploring urban spatial reorganization at large and transportation terminals as specific mobility facilities, in particular. In this last chapter of this part we will return to a focus on people, attempting to explore the potentially unrestricted global spatial or locational opportunities which daily virtual spatial mobilities may offer to individuals. Thus, we will first argue for a possibly new significance of geographical location in the information age, amounting to opportunity. Location as opportunity suggests that individuals can perform basic daily activities beyond their corporeal spatial anchoring as well as beyond the traditional locations of basic facilities for daily activities, such as stores, banks, colleges and work places. With the aid of well-developed and widely adopted virtual mobility media, location does not constitute anymore an absolute locational destiny as it used to be in the industrial age (Kellerman and Paradiso 2007).

In the previous chapters, notably in Chapter 7 devoted to virtual daily mobilities, we focused on the locationally unbound, daily virtual mobilities, offered by mobile broadband services operated through laptops and smartphones. The wide availability of location-free communications constitutes foremost a complete locational flexibility for the origin of calls and callers. Obviously, this flexibility can be used also for the reaching of any chosen personal call destinations, such as work places and social contacts, or for the retrieval of any type of information whether personal (such as banking services) or public (such as news or music). This chapter will focus on the unrestricted potential for the reaching and transmission of personal information implying locational flexibility of basic daily activities. Virtual mobility media at the service of individuals coupled with cyberspace web locations developed by companies and organizations may jointly bring about emerging opportunities for global choices of destinations for services which were once considered place-bound, located in the physical proximity of users (such as shopping and learning).

The chapter will consist of the following parts: first we will discuss transitions in the significance of location at large. This will be followed by short expositions on geographical location as destiny for individuals in the industrial age. Third, we will highlight a transitional period spanning the 1970s and 1980s, during which geographical location lost much of its absolute 'destiny'-like nature, thus becoming more of an anchor. These sections will basically constitute a synopsis based on literature reviews. In the fourth section we will attempt to present a conceptual framework for geographical location turning into opportunity for individuals in the information age.

We will elaborate, within the fourth section, on individuals' geographical locational opportunities in the information age, in which people are empowered by information and communications technologies. We will highlight online shopping, e-learning, home-based business/work, and social networking. These four activities may potentially reach global destination and distanciation as far as location of stores, universities, business clientele, and friends, respectively, are concerned, since shopping, learning, business/work, and social networking can be potentially performed at any distance from one's home assuming the use of virtual mobility media. As we will see, these potential opportunities have only partially been materialized by individuals, with wider adoption in North America, and a rather differentiated one in Europe. The examination of these four concrete uses of daily virtual mobility, online shopping, distance learning, home-based business/work, and social networking, is restricted by data availability, so that the chapter should be viewed mainly as suggesting a contemporary perspective on geographical location. Finally, in the concluding section of this chapter, we will elaborate on policy issues pertaining to the fostering of locational opportunities in the information age.

Another popular service-oriented Internet activity is e-government. However, this activity is irrelevant here when locational opportunities are examined, since governments do not provide for choice flexibility or locational opportunities as far as service providers are concerned. Thus, one cannot choose between local, regional and national governments opting, for example, to receive some governmental services from foreign governments or cities. In this respect, e-government is completely different from shopping, learning and working which became potentially, at least, location free from the perspective of service supply. On the other hand, however, e-government services permit a geographically unlimited access to them in that they can be reached by relevant citizens/residents from anywhere globally. This global access may sometimes require some technical manipulations as far as the use of domestic languages in foreign countries is concerned.

Location: From destiny to opportunity

"Geographical location is no longer destiny", declared Craig Barrett (2005), Chair of *Intel* at the Tunis WSIS (World Summit on the Information Society). Geographical location still maintains, though, even in the information age, its basic dimension as a physical anchor in space for activities, including those which are typical of the information society, such as special locational requirements for R&D activities in the information technology industry, and particularly those needed for the Internet industry (see Kellerman 2002). Hence, we intend to trace a possibly evolving change in the significance of geographical location for individuals from *destiny* in the industrial age to *opportunity* in the information age applying to individuals' lives in the information society. The examination of new flexibilities in locational opportunities for individuals may be viewed as a new context for the classical place utility approach (see e.g. Meier 1962).

Locational opportunity can be potentially considered as mature only in the information age. The novelties induced in the information age, and having significant implications for location, are not only related to the Internet as cyberspace and its reaching through communications media. It is related also to the complex interplay between traditional realities of service locations in geographic units, to new spaces of interaction and expansion for individual actions as well as to the embodiment of individuals' experience reaching a wider range of activity. In other words, locational opportunities potentially exist for global economic activity of individuals, but this is not to say, though, that societal traditions and governmental bureaucracies have ceased to slow down and control global activity by individuals.

From the perspective of individuals, they enjoy additional flexibilities for the location of both economic and social activities in the information age, implying global daily mobilities, but such enhanced flexibilities do not automatically call for actors to prefer the initiation of actions in places other than their residential ones. Thus, whereas destiny for geographical locations implied dictation in the past, opportunity for geographical locations constitutes a potential, and not necessarily yet a mature reality. The materialization of this new potential might take time (Kellerman 2006b). Proper policies undertaken by local, regional and national governments may contribute to beneficial disseminations of the opportunities for daily actions offered by the information age.

Locational destiny in the industrial age (1840–1970)

In the early industrial age, location for individuals constituted an absolute concept of destiny. It expressed rather restricted mechanical technology competences of individuals, yielding work and service locations at or near homes, employing animal-based traditional transportation systems. In this early industrial age individuals constituted foremost factors of production in virgin systems of rules related to land access and disposal, as well as to civilized settlements for workers. Individual mobility in the industrial age was generally linked to needs of localization of industrial activities. Everyday practice under traditional conditions was closely tied to physical-material conditions (Werlen 2005: 3).

Narrow spatial limitations, or 'embeddedness' in a spatial sense, lay in the technical standards of traditional transportation and communication media. "The predominance of walking and the limited significance of writing restricted social and cultural expressions to the local and regional level. Face to face interaction was almost the only possible situation for communication" (Werlen 2005: 3). This represents the village-system of territorialization, which was deeply, though gradually, changed by industrial activities which transformed the spatialization of human communities introducing the urban scene as the dominant location.

Early development of new transportation technologies from the mid-nineteenth century, notably trains, led to later constructs of more nuanced destiny concepts, given that the spatial extent of places, notably cities, expanded and permitted more movements about them. Newcomers to urban areas, mainly in the nineteenth century, found themselves often living under uncivilized life conditions, without health care systems and with low access to education (e.g. slums in English cities). Twentieth century Western Europe was typified by enhanced living conditions and equality, simultaneously with the emergence of civil societies, which grew out of the affirmation of nation-states, and later on bringing about the social mix which was produced by the two world wars, side by side with an accent on welfare-state systems. However, the progressive development of communications and transportation media, coupled with the emergence of mass markets did not yet change at that time the persistence of place relations for individuals, constituting place-tied communities, in which age, gender, and class mattered (Werlen 2005). Spatial flexibilities in locational opportunities were still limited to the metropolitan scale requiring corporeal mobility using trains, buses and cars for employment and service purchasing, etc.

Locational anchoring in the post-industrial age (1970s–1980s)

The two decades of the 1970s–80s were marked by the fragmented nature of the post-industrial age. The growth in the service sector and early developments of an information-based society implied, on the one hand, the emergence of wider individual enabling, and opportunities emerging in a richer frame of paths, and on the other hand, they brought about an increasing uncertainty for reversible conditions, notably as far as job access and maintenance was concerned. Some commentators went as far as arguing for 'the end of job' as one of the main future trends (Rifkin 1995, see also Bonss and Kesselring 2004, Kellerman 2000).

Individuals in post-modern society experienced new life conditions through the emerging common search for new experiences and interactions. Thus, for some commentators the sense of self was no more unique or authentic but continuously bargaining with external marketing messages and multiple relational situations: the sense of self seemed to evolve under a condition of 'multiple form individuals', who compared to individuals of the past, were culturally oriented to live the temporary conditions of the present rather than future building or past memory for the future (Gergen 1991, quoted in Rifkin 2000). Individuals seemed to increase their ability to reinvent themselves as network nodes, not merely attached to bounded locations. In EU countries these trends were accompanied by regional policies which introduced new potentials for local experiments in terms of networks of individuals and institutions (Kearns and Philo 1993, Paradiso 1999). The metaphor of *milieu innovateur* underlined the increasing power of highly skilled people organized in chains, which could have been enhanced by ICTs. However, emerging lifestyle fragmentation and diversification implied dualistic societal systems, in which individual actions were conditioned by the educational levels of individuals and resulting jobs and opportunities (Castells 1989), coupled with the increasing importance of marketing and consumerism, bringing some to suggest a 'fetishist' society in which the identity of individuals was more linked to object-possession than to personal connotation or moral values (Baudrillard 1988).

All these trends of growing personal flexibility in the new modernity paved the way for the later upcoming information age which offered people the personal technologies for growing mobilities for individual production work, consumption and personal growth.

Locational opportunities in the information age

The information age of the second modernity, maturing since the 1990s, has been mainly typified by a fast adoption of information technology in almost all aspects of human life and action, as we noted already in previous chapters. Virtual mobility has emerged as a new vibrant dimension of society expanding communications possibilities and permitting communications while on the go corporeally. Thus, the major changes in the significance of geographical location in this era relates to mobility. Information technologies and media, mainly the Internet and mobile telephony, have become rapidly ubiquitously available in developed countries, as well as growingly in the developing world. Mobility at large has been a major characteristic of modernity: "modern society is a society on the move" (Lash and Urry 1994: 252): humans being constantly on the move in both society and space. The wide availability and growing significance of mobilities has potentially brought about expanding locational flexibilities and opportunities for individuals.

For locational opportunity, enhanced virtual mobility implies self-extension and, thus, more opportunities for mobile persons at the two ends or terminals of any act of communications/mobility. In terms of points of origin, communications media permit one to reach the same people and the same information, or the same destinations, from wherever one is located, or from any point of origin. But more importantly from the perspective of locational opportunities, is the following: the distanciation or the increasing geographical spread of potential destinations for person-to-person communications, the retrieval of information, and the ability to work and to buy services globally, have widened locational opportunities and individuals' action spaces (see Giddens 1990). Cyberspace has not merely been a new world of information, but it is also a dense concept of converging technological environments, human minds, personal motivations, and a generation source for all kinds of human artifacts (Dodge and Kitchin 2001).

The three major communications devices, fixed and mobile telephones and the Internet, share self-propelled operations by users for the transmission and receipt of information. The contemporary scene of the information age presents much choice among media for personal mobility, for vocal communications (via telephones, mobile phones and VoIP), as well as for written communications (through e-mail, SMS, and fax). These media of personal mobility, thus, differ from earlier information transmission services which depend on intermediary operating agents, notably the telegraph and postal services. These latter public virtual mobility media provide a rather lagged interaction among users as compared to the simultaneous interaction provided by telephone and Internet services.

As we noticed already, reaching new opportunities has become even more flexible through the possibility for two people to communicate, while both being on the road, so that both parties perform simultaneous personal physical and virtual mobilities. Other new possibilities include a person on the road communicating with information sources over the Internet Web, with changing virtual locations ('hosting') for these websites as the person moves, but without any ramifications to the user's ability to access these websites when the locations of these information sources change. These unprecedented forms and patterns of personal virtual mobility may imply an adaptation of mobility media to given spatial structures rather than facilitating their change (see Chapter 9). In other words, new sophisticated virtual mobilities permit us to be in touch without any regard to the location of the cities or countries from which we communicate, nor with regard to the internal spatial structures of cities from which we call.

Reaching new opportunities via information technology may be viewed as a sequential process which involves several phases, detailed elsewhere (Kellerman 2006a), and illustrated in Figure 11.1. First, a changing balance between human fixity and mobility through the very possible use of electronic communications means, by accentuating mobility at the expense of but not as a nullifying of fixity. In other words, we become highly mobile virtually and corporeally but we still need fixed places for homes, etc. The operation of contemporary personal communications media (the fixed and mobile telephones and the Internet) is assumed to be rather convenient and user-friendly, and thus barrier-free. Once communications is initiated, the speed of reach of other places and people is normally instant, and the constant enhancement of communications speed is an expectation of contemporary media users. Finally, the convenience and speed of communications result in the expansion of extensibility and accessibility of users reaching eventually new or veteran opportunities (see Janelle 1973, Adams 1995, Kwan 2001). Repeated reaching out to such opportunities may further change the balance between services with fixed locations within easy physical reach' on the one hand, and virtual services provided through the Internet or services located anywhere and reached through the Internet, on the other, in favor of the latter. New locational opportunities may enhance personal fulfillment through virtual cyberspace by making it possible for people to communicate freely via e-mail, blogs, chat networks, etc. These channels provide for free expression and exchanges even in countries of limited democracy, with a special significance for women for whom the Internet is sometimes an alternative to other forms of reaching out. Locational opportunity applies, thus, not only to the potentially global provision of services but to the expansion of social ties as well.

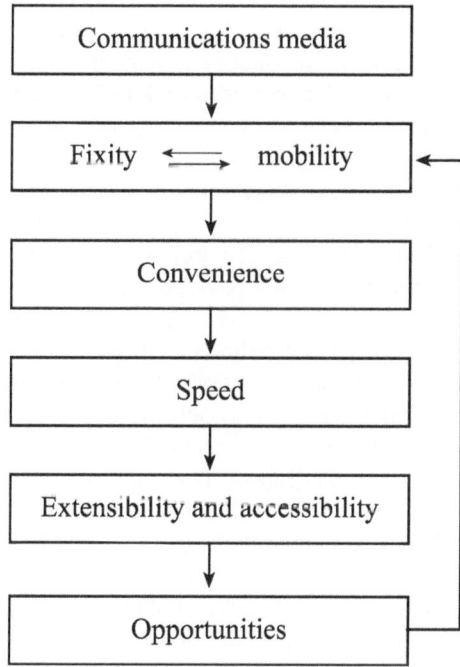

Figure 11.1 From communications to opportunities

Online shopping

Our discussion here will focus on what for merchants is called B2C (business to customers) e-commerce which for customers is usually termed *online shopping*. We will briefly examine a possible emergence of geographical opportunities for individual consumers to shop beyond their normal daily physical reach *vis-à-vis* online shopping, and at two potential geographical levels, domestic and international. The data which will be presented and interpreted are mostly at the national level, as systematic data on online shopping at the regional and local scales, are not yet available. Furthermore, our discussion here will focus on 'shopping' in its rather traditional sense of the purchase of products and not on the purchase of services, such as flight and hotel reservations, which are increasingly made through the Internet.

The Internet permits us to perform virtually all the phases of physical shopping, except for touching and trying: search for products; search for vendors; viewing of products and reading of information on them; price comparisons; and eventually their very purchasing. Shopping has, thus, become geographically extended into a potentially global 'opportunity'. On the other hand, one has to bear in mind the *fragmentation of activity*, calling for some, and if only basic, continued physical shopping (Couclelis 2004; see also Schwanen *et al*. 2008). Do people actually make use of this global opportunity for shopping without any geographical restrictions of place and distance, or do they prefer to stick to the traditional way of shopping in physically accessed stores? This question is even more interesting when online shopping is divided between domestic websites ('closer' ones, geographically and culturally) and foreign, more 'remote' ones.

For many contemporary customers, shopping may involve a mixture of online activities and physical ones. Thus, one may browse for products, makers, sellers and prices and eventually shop in a physical store or *vice versa*. The affordance of the Internet for e-commerce is, therefore, in many cases, partial (Schwanen *et al*. 2008). We will focus here on online purchasing, as this concluding phase of shopping implies domestic or international transfer of funds, as well as remote domestic or international guarantee by sellers. Our focus will neither be on the aspects of financial volume of spending by individuals, nor on the financial extent of B2C e-commerce, and other aspects (see e.g. Kellerman 2002: 127-33, Visser and Lanzendorf 2004). Rather, we will concentrate on the behavior of customers, in order to see whether the opportunity to shop indifferently from a locational perspective is really materialized, or alternatively the physical locational 'destiny' for shopping in physical stores still governs shopping behavior.

As far as online shopping in general (domestically and in foreign countries) is concerned, measured by the percentage of the total population shopping online, it seems that the use of this medium has become quite popular in North America and less so in Europe. Thus, already by 2003 some 40 percent of Canadian households made online purchases. However, Ontario accounted for almost one half of total online spending (Statistics Canada 2004). In the US, in 2001, some 21 percent of the total population purchased online (NTIA 2002). However, the share of e-retail sales out of the total retail sales in that year was only 1.1 percent, rising to 2.4 percent in 2005 (Malecki and Moriset 2008), and to 3.6 percent in 2008 (US Bureau of the Census 2011). For 2008 the leading countries in the percentage of people who shopped online (both domestically and internationally) out of those having access to the Internet were: South Korea (97 percent); UK (97 percent); Germany (97 percent); and US (94 percent). In the EU in general, in 2005–2006, only 23 percent of the total population did so, and this percentage was equal to the share of the population purchasing through the more traditional, but still permitting shopping at a distance, postal service! (European Commission 2006). Four years later, in 2010, this rate increased to 33. The EU, though, is highly diversified, including countries with high population percentages of domestic online shopping in 2010, led by Sweden (54 percent, the Netherlands (52 percent), and Germany (47 percent), going all the way down to 7 percent in Malta. Obviously, the data for 2005–2006 present more modest rates (Table 11.1).

Table 11.1 Percentage population shopping online in EU countries and/or domestically, 2005–2006 and 2010

Country	Shopping in EU countries 2005–6	Shopping domestically 2005–6	Shopping in EU countries 2010	Shopping domestically 2010
Luxembourg	28	7	38	12
Denmark	19	46	24	32
Austria	18	21	30	32
Netherlands	15	46	12	52
Sweden	14	45	13	54
Finland	13	34	18	38
Ireland	12	19	34	26
Belgium	12	17	11	19
Malta	11	4	39	7
UK	7	41	9	53
France	7	26	9	38
Germany	4	30	6	47
Estonia	4	17	8	21
Italy	4	11	4	13
Slovenia	4	10	8	23
Spain	4	9	6	21
Cyprus	4	0	22	11
Czech Rep	3	21	4	38
Latvia	3	11	7	17
Poland	2	16	3	35
Portugal	2	3	5	12
Greece	1	2	8	12
Slovakia	1	5	18	30
Hungary	1	8	4	23
Lithuania	1	4	7	17
EU general	**6**	**23**	**7**	**33**

Sources: Data: European Commission, 2006 (illustrations D46 and QB1.1) and 2011 (p. 15).

Purchasing abroad may be considered as an advanced form of domestic online shopping, as it may involve money exchange, trusting a foreign company, as well as overcoming an image of remoteness. Here too, there was found a major difference between North America and the EU. Whereas in Canada in 2003, one third of online purchases were made through foreign websites (Statistics Canada 2004), probably mainly American ones, in the EU in 2005–2006, the percentage of purchases made in other EU countries stood at only 6 percent, increasing only modestly to 7 percent in 2010 (see also Aoyama 2003). This was despite the convenience of possibly using the same currency, the Euro, in most purchases. The growth trend for cross-border online shopping in the EU was suggested at the time to be generally high, with several countries doubling the percentage of population use from 2002 to 2006 (European Commission 2006). Between 2006 and 2010, though, alongside impressive growth in cross-border shopping in small countries (e.g. Ireland, Cyprus and Malta), other countries presented slight declines (e.g. Netherlands, Sweden and Belgium. As things turned out, the EU percentage of homes connected to the Internet grew from the 2005 rate of 44 percent to 60 percent in 2008 (Eurostat 2008), but the percentage of those who purchased abroad through the Internet grew very modestly from 6 percent in 2005–2006 to just 7 percent in 2008 and 2010. It was though for domestic online shopping to have grown much faster between 2005–2006 and 2008, from 23 percent to 30 percent, and again to 33 percent in 2010, attesting to the still existing apprehension of international shopping online even in other EU countries (Commission of the European Communities 2009, European Commission 2011).

The European Commission has carried out surveys of some 25,000 citizens in its 25-27 member-countries on a biannual basis, which referred to both domestic and foreign (in other EU countries only) online shopping (European Commission 2006, 2010) (Table 11.1). The country data for these two geographic extents of shopping do not permit formal statistical analysis because of the rather small samples used in countries with low Internet penetration rates and low rates for online shopping. Still, however, when ordering the countries by decreasing magnitude of foreign purchases, two factors seem to govern European online shopping behavior. First, country size matters. Citizens of small countries tend to perform more foreign shopping then those of larger ones. Thus, Luxembourg led the list of foreign shopping in 2006 with a percentage (28) which was four times larger than its percentage domestic online shoppers (7), and Malta did so in 2010. They were followed by various other small countries. This same factor applies also to small countries with modest values of online shopping. Thus, Cyprus had no domestic online shopping at all in 2006, whereas the Maltese population tended over 5.5 times more to purchase internationally than domestically in 2010 (39 percent and 7 percent respectively). The opposite trend was true for three of the larger European countries, namely more domestic online shopping than foreign: UK (41 percent for domestic shopping versus 7 percent abroad in 2006, and 53 percent respectively for 2010); France (26 percent and 7 percent for 2006, and 38 and 9 for 2010 respectively); and Germany (30 percent and

4 percent for 2006, and 47 and 6 percent for 2010 respectively). Online domestic shopping opportunities are larger in bigger countries, so that the very availability of foreign online shopping provides an opportunity for shoppers in smaller countries more than for shoppers in big ones. The importance of country size is of interest since previous studies of international B2C e-commerce identified Internet infrastructure, economic development and cultural factors as key aspects for differences in magnitude (Hwang *et al.* 2006, Kshetri 2001).

The second factor for international online shopping in the EU is national Internet penetration rates. The Scandinavian countries have led the adoption of Internet and telecommunications media at large since the introduction of the telephone (Kellerman 1999). They also lead in both foreign and domestic online shopping. The Internet has, thus, changed, to some degree, the shopping habits of Scandinavians as well as those of residents of other small and well-developed EU countries.

The overall picture of the rather modest percentages of population performing online shopping seems to suggest that geographical location as 'destiny' is still there, at least partially, as far as shopping is concerned. This 'destiny' of geographical location, pertaining to shopping by individuals, is there because of shoppers' attitudes. Habits, trust, and uncertainties are main obstacles (European Commission 2002), but the willingness to touch and to try merchandise, as well as its purchase tend to constitute also a kind of entertainment, and are additional factors for slowing down a wide adoption of online shopping. On the other hand, e-commerce may also turn into a form of empowerment for women, notably in developing countries. On the supply side, merchants have to invest more in foreign countries in order to attract customers, foremost over the Internet, targeting mainly the younger generations for whom cross-border activities at large might be more liberated. Needless to say that websites have to be multilingual in order to facilitate shopping, notably if there evolves some competition between international and domestic websites.

E-learning

Another opportunity becoming potentially available to individuals in the information age is e-learning through the Internet, both in domestic institutes or through cross-border education. One may study via the Internet at various levels. Tacit knowledge may be acquired and thus enrich the traditional frontal study through Web browsing, or via virtual consultations with remote colleagues. Codified knowledge, on the other hand, may be achieved through formal courses or through full academic degree studies. Whereas the first two options are informal and may range in volume from time to time, formal degree studies are similar to e-commerce in that a product, knowledge, can be bought online. There are, however, various differences between the sale of products and services (such as airline tickets) over the Internet, on the one hand, and the provision of formal codified knowledge through the Web, which may bring about differing levels of their adoption. Some of these differences are highlighted in the following paragraphs.

From the supply side, language is a much more crucial element in e-learning than in e-commerce, by the very nature of learning. Standards is another and even more crucial dimension of formal study when compared to internationally recognized quality standards for products and services (e.g. ISO 9000), since a domestic or cross-border academic degree has to be recognized by national higher education councils, as well as by potential employers. At the international level of high-education standards for an opening and globalizing world, several efforts have been made to assure such standardization, e.g. the EU Bologna agreements, as well as global GATS (General Agreements on Trade in Services) agreements (Knight 2006).

From the demand side, the purchase of an online degree program implies a prolonged purchase process spanning several years, as compared to a few minutes when buying other products and services online. It further requires a much more extensive investment by the customer, in terms of financial costs, time, and intellectual efforts. This latter intellectual effort may make online studies inferior to face-to-face ones, from the perspectives of both students and universities and colleges. As far as product durability is concerned, an academic degree constitutes a lifelong product as compared to disposability or limited lifespan of other products and services.

All these reasons and others have turned online formal academic education into a widely debated issue, but still relatively modestly used, even as compared to e-commerce (see e.g. Breton and Lambert 2003, and a review in Huh 2006). The *eUSER Population Survey 2005* carried out by the EU in ten member countries did not inquire about online full-degree studies. It yielded low results for the more modest option of online studies of at least one course (Table 11.2). In terms of the percentage of e-learners out of the total adult learners (and not out of the total population), the results ranged from the rather low 3.3 percent in Germany, which ranked much higher in domestic e-commerce, to the leading UK (10.7 percent) and Ireland (10.0 percent). The relatively higher values for the UK and Ireland may be related to the wider availability of study opportunities in English, as well as to the impact of the British Open University. The need to overcome traditions, habits and conventions by potential students and the hesitation of many universities and colleges to fully teach over the Internet require careful attention in the tracing of possible future developments of study opportunities over the Internet.

Table 11.2 Online course learners as percentage of adult learners, 2005

Country	%
UK	10.7
Ireland	10.0
Denmark	7.5
Poland	6.5
France	6.0
Slovenia	5.2
Czech Republic	4.8
Italy	4.7
Hungary	3.9
Germany	3.3
Overall sample	**6.4**

Data source: EU 2005.

In the US, some 56 percent of the degree-granting institutions offered distance courses in 2000–2001, but not necessarily full-degree studies. These offerings ranged from just 16 percent of the private two-year colleges, to 90 percent of the public two-year colleges, and from 40 percent of the private four-year institutions to 89 percent of the public four-year institutions (US Bureau of the Census 2006a). The major differences between the two sectors of universities and colleges reflect academic differences regarding e-learning and responsibility towards potential clientele, with the two-year colleges more inclined to foster higher education in peripheral regions and to cater for working students who need more flexible learning times. The data reported by the US Bureau of the Census do not permit to assess the percentage of students who take at least one distance course, but the data on the offerings permit to assume that this form of study is more popular in peripheral and low-income segments of the population, so that the Internet provides an opportunity for students who otherwise would not have been able to study for an academic degree.

Home-based business/work

Home-based work, or as it has often been termed as *telework, telecommuting,* or *telehomework,* has probably been the first and continuously discussed potential locational opportunity for individuals equipped with ICTs (see Kellerman 1984). For Halford (2005) work at home implies the spatial *relocation* of work coupled with its *dislocation* into cyberspace. Work at home, fully or partially, became potentially possible as far back as the 1970s following the diffusion of the *Telnet* technology for remote access to mainframe computers, using computer terminals and later on PCs. The introduction of the Internet and its extremely wide options for information transmission some twenty years later, in the mid-1990s, should have made work from home become a standard form of work following the 20-year experience, at least a potential one, of partial possibilities for remote work. However, as things have turned out, work at home as a partial or exclusive location for work has not become a favorite option, neither for employers, nor for employees or for the self-employed. Employers preferred on-site supervision as well as face-to-face staff meetings on a regular basis, and employees preferred to see colleagues, be involved with goings-on in their offices, and women in particular tended to prefer a geographical separation between home and work duties (see Halford 2005, Hislop and Axtell 2007, and Felstead *et al.* 2005 for reviews). Work at home has normally not implied a higher quality of life for workers, though productivity was reported to be higher at home (Halford 2005).

Home-based workers may be classified into three groups: partial or hybrid; full-time and mobile workers who work at various places such as at clients' facilities (see Malecki and Moriset 2008). Of these three groups people who work wholly from home are of special interest, since such workers are similar to those who perform full purchase of products and services online or those who study for an academic degree fully through cyberspace. Even more interesting are the self-employed who work fully from home, mobilizing their clients through the Internet, and performing their work through the Internet as well, e.g. text editors and translators or consultants.

For the year 2000 a low 0.6 percent was estimated for employees who worked wholly from home out of the total employees in European countries (Bates and Huws 2002; see also Helminen and Ristimäki 2007). In 2005 the percentage of those involved in telework 'almost all of the time' in EU-27 countries rose to 1.7, and an additional 7.0 percent were involved in telework at least 'a quarter of the time' (Welz and Wolf 2010). For the UK in 2002 it was estimated that 2.4 percent worked mainly, but not fully, from home (Hislop and Axtell 2007) rising to 2.5 in 2005 (Welz and Wolf 2010). In the US, which was the pioneering society in the adoption of computers for both work and homes, in 2008 only 1 percent of employers allowed all or most of their employees to work some regular paid hours at home on a regular base (US Bureau of the Census 2010b). In 2005 some 5.6 percent of US workers worked exclusively from home. However this percentage included farmers, construction workers and manufacturers. Thus, only some 1.7 percent of US workers performed professional and related services from home (see US Bureau of the Census SIPP 2006b).

The relationships between locational opportunities for work from home, on the one hand, and the remote provision of services and studying from home, on the other, require some separate attention. It might well be that the extensive exposure to computers and working with computers at normative work places outside of home brings about the use of computers more extensively at home as well, albeit for non-work activities such as the provision of services, shopping and study rather than acquiring these services, like work, also outside of home. Such a relationship could potentially also be true the other way around, namely that people who work at home would prefer to shop and study away from home. However, there might still be a third group of people, who would opt to do everything possible through the Internet, including working, shopping and studying. As things look like in the early 2010s, there is a growing tendency to make use of the Internet for the provision of many services, but not for work from home even if percentages of workers at home tend to grow modestly.

Social networking

The three locational opportunities which are potentially available to individuals and which were presented so far in this chapter are of an economic nature, involving the remote purchasing of products and services, including knowledge, as well as the selling of labor and the search for business. The locational opportunity for social networking is different in its being purely social, referring to the ability of people who are connected to the Internet to contact other persons who subscribe to social networking systems, notably *Facebook* and *Twitter*, and who are located anywhere globally (see Chapter 1). Side-by-side with social networking, business and work networks have emerged as well, whether through access-restricted Intranets or whether through any other more open systems (see Castells 2000, 2001). These latter business/work networks are not of our interest here due to their very nature of business objected networks, attractive or available only to persons involved in a certain field or job.

Social networking nested within the Internet even before its inception as a wide and open access system in the mid-1990s. Thus, some global networks developed initially around a physical location (e.g. the San Francisco-based *WELL* network) (see Rheingold 1993), whereas others, such as *MOOs*, were organized around a symbolic city, implying centrality and agglomeration in the volume and intensity of communications to specific 'rooms', 'buildings', or 'neighborhoods' (see Schrag 1994).

A second generation of social networking, becoming popular from the mid-1990s, consisted of on-line exchanges (e.g. via ICQ or MSN) and blogs. Data for the latter for October 2005 estimated an existence of about 100 million blogs worldwide, including a high percentage of spam. International differences in their country distribution existed within both the developed and developing worlds. The US dominated the scene with some 30-50 million blogs, while neighboring Canada had only approximately 700,000. Similar gaps were witnessed in Europe as well. Thus, France had some 3.5 million blogs, whereas Germany had only 300,000. By the same token, Spain and Poland had 1.5 million each, whereas in Italy there were only 250,000. Leading industrializing countries presented similar trends of high international differences, with China having 6 million blogs, while India had only 100,000 (*The Blog Herald* 2005).

Social networking has frequently led to significant changes at both the personal and the societal levels. Personally, relating to personal relations, Ben-Ze'ev (2004) presented the evolution of romantic and sexual relations all the way from initial electronically-written contacts through e-mail, real-time exchanges, blogs, and SMSs, followed by video conversations, to cybersex activities. At the societal level, the social networking systems served as major tools for the provoking and sustaining of political unrest and revolution in 2011 in Middle Eastern and North African countries.

The networking dimension of the Internet has been accentuated in a thrid phase of social networking opportunities since the early 2000s with the emergence of Web 2.0, maturing in the late 2000s with a focus on social networking through swiftly adopted networks, led by *Facebook*, *Twitter*, *MySpace*, etc. *Facebook* has come to constitute the major competitor for *Google* on dominance in the Internet, with the latter specializing in Web 1.0 applications, mostly search engines, and *Facebook* specializing in Web 2.0 applications, notably social networking. *Facebook* was established in 2004. As we mentioned already, on average some 900,000 blog posts were created daily in 2009, with some 200 million people or 13 percent of Internet users worldwide using *Facebook* actively in 2009, and one half of them did so at least once a day (Scherr Technology 2009). Just less than two years later, in January 2011 *Facebook* claimed to have over 600 million users (*Business Insider* 2011) out of a total of c. 2 billion Internet users worldwide (*eMarketer* 2011), or about one-third of Internet users globally made use of *Facebook* by then! We can assume that by adding the users of other networks, such as *Twitter* and *MySpace*, at least three-quarters of Internet users were engaged in social networking over the net in early 2011. Thus, the barrier-free, constraint-free (in most countries) and cost-free locational opportunity of social networking has exhibited tremendously fast adoption rates. However, the very use of social networking over the Internet does not automatically imply globally stretched networks. Facebook has become, for example, a framework for virtual interaction among classmates for school kids whose location may not stretch beyond a single neighborhood.

Table 11.3 presents a country breakdown of *Facebook* subscribers for the 30 leading countries by the end of 2010 as compared to Internet users some six months earlier. The fast adoption of *Facebook* makes this gap of only six months most significant as shown, for example, by the data for Indonesia. Looking at the percentage of *Facebook* subscribers among Internet users, the five leading countries among the 30 leading by the number of *Facebook* subscribers are Indonesia (100 percent?); Chile (90.6 percent); Venezuela (81.2 percent); Hong Kong (75.3 percent); and Turkey (69.0 percent). Obviously there might be numerous other countries with lower numbers of *Facebook* subscribers but still presenting high penetration rates. The five countries among the leading 30 with the lowest penetration rates are Brazil (4.5 percent); Poland (20.2 percent); Germany (21.0 percent); India (21.3 percent); and Netherlands (23.0 percent). Interestingly enough some three quarters of the global subscribers to *Facebook* were outside the US (in which the system was first introduced in 2004) just six years before the 2010 data. One cannot classify the 2010 *Facebook* penetration rates by the usual diffusion process for communications media, beginning in the US and in Scandinavian countries and then moving through the developed countries into the developing ones (see Kellerman 1999). Thus, the diffusion of no-cost social networking, independent of levels of economic development, representing the seeking for merely social relations, has its own rules of diffusion and penetration, probably related to national cultures of social interaction at large, and mediated one in particular.

A darker side of networking is surveillance, the wide discussion of which is beyond the scope of this volume (see e.g. Murakami Wood and Graham 2006). It would suffice here to note that despite the seemingly autonomous nature of personal virtual mobilities at large, communications may undergo surveillance and be turned into recorded mobility. Internet use by surfers may be recorded by commercial companies in order to channel advertisements to target potential clientele, mobile telephone users may be located through GPS and other technologies, and telephone calls may be illegally recorded as well. Though these surveillance activities present diversified motivations they imply a risk, notably for those engaged in intimate social networking. Paradoxically, surveillance may hurt most notably those people who are involved in social networking activities who wish to express through its use their autonomy of operations.

Table 11.3 *Facebook* users as percentage of Internet users

Country	No. of *Facebook* users, Dec. 2010	No. of Internet users, June 2010	Percentage *Facebook* users
US	145,749,580	239,232,863	60.9
Indonesia	32,129,460	30,000,000	100?
UK	28,661,600	51,442,100	55.7
Turkey	24,163,600	35,000,000	69.0
France	20,469,420	44,625,300	45.9
Philippines	18,901,900	29,700,000	63.6
Mexico	18,488,700	30,600,000	60.4
Italy	17,812,800	30,026,400	59.3
India	17,288,900	81,000,000	21.3
Canada	17,288,620	26,224,900	65.9
Germany	13,678,200	65,123,800	21.0
Argentina	12,359,260	26,614,813	46.4
Spain	12,235,080	29,093,984	42.1
Colombia	11,665,860	21,529,415	54.2
Australia	9,661,720	17,033,826 (Sep. 2010)	56.7
Malaysia	9,544,580	16,902,600	56.5
Brazil	8,821,880	75,943,600	4.5
Taiwan	8,752,640	16,130,000	54.3
Chile	7,586,060	8,369,036	90.6
Venezuela	7,552,760	9,306,916	81.2
Thailand	6,732,780	17,486,400	38.5
Egypt	4,634,300	17,060,000	27.2
Poland	4,540,320	22,450,600	20.2
Sweden	4,042,260	8,397,900	48.1
Peru	3,888,560	8,084,900	48.1
Belgium	3,850,300	8,113,200	47.5
Hong Kong	3,673,580	4,878,713	75.3
South Africa	3,422,920	5,300,000	64.6
Netherlands	3,417,540	14,872,200	23.0
Saudi Arabia	3,274,460	9,800,000	33.4

Sources: Data: *Facebook* users: Nick Burcher 2011. Internet users: Internet world stats 2011.

Conclusion

Space and distance were argued time and again to constitute 'tyranny' for people and societies functioning under past technologies (e.g. Toffler 1980, Duranton 1999, Blainey 1966, Mitchell 1995). However, for the contemporary information age, Kauffmann (2002: 8) stated that "space is [thus] undefined and open. It is a set of opportunities in perpetual reorganization". This change from tyranny, or our preferred term of destiny, to opportunity, regarding location, was the focus of our elaborations in this chapter. As far as individuals are concerned, flexible locational opportunities, for instance online shopping, e-learning, and home-based business/ work, have become widely available, but only partially, though growingly, adopted. This tendency reflects long established habits for local and domestic geographical anchoring of basic human activities. As Hanson (1998: 248) colorfully stated: "Life on the ground is surely more complex than life entirely off the road; but then life entirely off the road is an oxymoron". However, as we noted, locational opportunities for social networking have been widely and swiftly adopted, presenting a more open approach to wide and flexible socialization as compared to slower and more conservative changes in work, shopping and learning habits.

The contemporary trends shown in this chapter, for potentially flexible locations of activities of individuals in the information age, strongly call for public policy and planning to foster locational opportunities for individuals. This need rises notably because of the slow pace of 'automatic' adoption of opportunities by individuals without institutional intervention. It would suffice here to make some general remarks on approaches to the possible roles for public policy and planning. Genuine opportunities for the enabling of individuals have to be sought by policy makers, given that contemporary individuals are engaged and embedded simultaneously within virtual and physical mobilities and fixities. Policies in this regard have to pay special attention to the removal of obstacles for locationally more flexible activities by individuals in both developed and developing countries. Such policies have to be sensitive to nuances among cities and regions, in both developed and developing countries, regarding local cultural, political and regulative systems. Examples in this regard are customs policies permitting personal imports, clear value-added taxation regulations, and the examination of academic degrees and credits offered by foreign universities through e-learning.

The development and implementation of policies aimed at the fostering of opportunities for individuals depends on governance. It may possibly be dialectical with the promotion of opportunities for the development of geographical regions. On the one hand, the availability of additional locational opportunities for individuals using information technologies implies a wide geographical expansion of locational opportunities, beyond the boundaries of their residential geographical area, and at the expense of local commerce and services. This expansion of opportunities may take place simultaneously with efforts made by local and regional authorities working hard for the development of these same regions. This dialectic presents one of the major axes of the contemporary information age.

Local residents of any region may act (e.g. shop) outside their area of residence using information technologies while people from other areas, using the same technologies, may be attracted to act (e.g. shop) in their region, if the proper context for mutual shopping in other regions exits or is promoted. Places which are able to gain 'incoming activity' more than 'outgoing activity', or at least keep the two balanced, will win the game in the contemporary open, globalized, and complex current world of action, driven by instantaneous information flows of daily spatial mobilities by individuals.

Chapter 12
Conclusion

This concluding chapter will attempt to highlight several elements in ways of summary and conclusion for the previous chapters, as well as in ways of putting forward some ideas for future study. Specifically we will first present a summary of the book by suggesting features of daily spatial mobilities based on Chapter 1, followed by a discussion of dimensions of daily spatial mobilities which will summarize Chapters 2–11. This will be followed by concluding treatments of each of the basic notions comprising together the very concept of daily spatial mobilities: a comparison between daily and non-daily mobilities; daily mobilities and space; and the mobile agent. We will conclude this chapter with an elaboration of management aspects for daily spatial mobilities.

Book summary

Features of daily spatial mobilities

Daily spatial mobilities were introduced in this volume as routine, two-way displacements, as compared to the non-routine two-way spatial mobility of tourism and the non-routine one-way spatial mobility of migration. If mobility at large, and technology-based one in particular, constitute a major feature of contemporary life, then daily spatial mobilities serve as their routine expression, experienced continuously and intensively by individuals, as compared to the non-daily mobilities of tourism and migration which are experienced either seasonally, occasionally or annually (tourism), or seldom or never (migration).

In Chapter 1 we noticed several key terms for the study of daily spatial mobilities: information society; globalization; space of flows; networking; fixity; directionality; circularity; speed; extensibility; accessibility; time-space compression and distanciation. These key terms can be tied together in order for them to provide a framework for daily spatial mobilities in the following way.

Within information society each human being may be considered as a node within the global information/communications system. Thus, daily spatial mobilities are contextualized within globalization in both the production and consumption of mobilities. Virtual global mobility through the Internet is anchored within the space of flows, whereas routine global corporeal mobility is anchored in those jobs which require global physical mobility. Much of contemporary daily communications is performed within networks, both at work and in social-personal communications. Despite of these growing mobilities, fixity is still of significance in the creation of mobility and *vice versa*. Movements in daily mobilities may be typified and assessed by their directionality and circularity, as well as through their speed which has become an important social value. The spatial extent of people's activities, assuming that people are equipped with contemporary mobility technologies and able to use them, has expanded. This is expressed through enhanced extensibility of users of technology-based mobility media and their wider accessibility to places. Given the nature of mobility technologies they imply growing time-space compression and distanciation. Thus, there has emerged a growing flexibility for individuals as far as movements, and action and activity spaces are concerned.

The common thread among these terms is that they may serve students of daily spatial mobilities in their analyses of mobility. For mobile agents, on the other hand, some other, but related, features are prominent in their performance and experience of contemporary mobility, as we saw in the previous chapters: flexibility (in the choice of media); variety (of media available for both corporeal and virtual mobilities); availability (always ready for use personal mobility media, notably private cars and media for virtual mobility); totality (the ability to transmit and receive all types of information virtually); and convenience (easy to operate media).

These features may bring about a vicious cycle of mobile agents continuously expecting more enhanced features and thereby causing themselves to become even more mobile. The industry of mobility media has a basic profit interest in technological advance of their products, so that the increasing sophistication of products and responds to demand may create new rounds of demands and supplies, through a continuous introduction of new models of media which may bring about even more extensive mobility. While this type of circular developments fits capitalist economies in developed countries, early signs of similar processes emerge also in developing countries in which the market for simple mobile phones and cheap laptops is on the rise, coupled already with information systems which were developed for mobile phones which serve rural populations (such as the Nokia $1.5 per month information system).

Dimensions of daily spatial mobilities

Chapters 2–11 portrayed a number of dimensions for daily spatial mobilities which we will summarize in this section. Following the introductory discussions of mobilities at large in Chapter 1, we presented a variety of psychological, geographical, economic, political and social motivations for daily spatial mobilities. With these motivations in mind we moved to explorations of potential mobilities and the very functioning and significance of daily spatial mobilities through their various modes of operation (terrestrial, virtual and aerial). We then turned to the examination of a variety of geographical implications of daily spatial mobilities, focusing mainly on spatial reorganization, transportation terminals and widening locational opportunities for numerous daily tasks (shopping, study, work and networking).

Mobility has been viewed as stemming from a number of psychological-social motives, consisting of proximity, locomotion, and curiosity, all of which were interpreted as bringing about primary demand for daily mobility. On the other hand, we noted that mobility may be viewed as derived demand as well, generated by numerous sources, and leading to attraction to specific people, places, events and information. Direct sources for mobility as derived demand are presented by people's activities at locations other than their homes, requiring them to move from their homes, as well as from other sites such as their working places, to a variety of destinations, either physically or virtually. Indirectly, a number of political and economic forces may determine or influence the production of mobilities. First are the very extensive mobilities of certain social sectors and their related power, producing immobilities of others for the sake of the functioning of their own mobilities. Second is urban planning which may give preference to specific forms of mobility, and third are economic forces maintaining the sustainability of cities through the provision of proximity for businesses and business people.

People's drives and incentives to move may lead to numerous patterns of mobility, of which personal mobility is most developed and preferred in contemporary societies. Personal mobility, in the form of individuals moving themselves by themselves through driving or calling, may be considered as personal autonomy: it constitutes a specific condition within a wider societal freedom to move. However, being free to move does not necessarily imply full or partial autonomous mobility, since the freedom to move may be constrained by a variety of both societal norms and personal tendencies. The attainment of personal autonomy through personal mobility involves a heavy use of mobility technologies, so that personal autonomy, as a social value, is not only expressed spatially through people on the move, but it is also embedded in spatial infrastructures of roads and cables, and even more so it is expressed in mobile artifacts like automobiles, laptops and mobile phones. When such vehicles/appliances stand idle they may still tell, by their very nature, that they permit their self-operation by their users, and, thus, of the personal mobility and personal autonomy of their owners/users. Since cars can be easily started and communications appliances easily carried by users, personal autonomy through personal mobility is an integral experience of contemporary mobile social life for all, either actively by users of personal mobility media or passively by temporary non-users.

The tendency of humans to be mobile in general and the execution of particular daily movements at specific times and places are subject to the potential mobilities of mobility agents. The concept of potential mobility was proposed as expanding the notion of motility, presenting the accumulation of mobility needs by individuals, their access to mobility media and their competences to make use of them. These elements may lead to an appropriation process which may result in several modes of mobilities, such as immobility, accessibility to other people's mobility or passive mobility, as well as several forms of active mobility, mainly movement *per se* and the potential exchange of one's mobility with those of other mobile agents, turning mobility into a form of capital.

Daily mobility may be performed terrestrially, virtually and aerially. These mobility types may be considered as a single entity despite the wide-ranging differences between these three major mobility types. This claim is based on the growing conceptualization of mobility as a single social structure. It is further based on several convergences among these three mobility types, as well as on the central role played by IT in all the three of them. Furthermore, all three mobility types, whether looked upon from the perspective of moving people, or when viewed from the perspective of mobility systems, share a basic model of mobility cycles: 'push' and 'pull' effects of mobility motivation, followed by the process chain of IT→origin→route→control→destination. This basic model is built into the very process of movement, and was relevant in the past to ancient mobility means such as carts and horses or mail pigeons, obviously in a simpler way without IT and with only a minimal or non-existent control of traffic. However, a conceptual change has typified contemporary technology-based mobilities, beyond just the tremendous increase in the speed of movements, namely the ability to transmit information through vehicle-free virtual mobility media. Messengers and postal systems have been replaced by a conversion of information of all types into digital bits transmitted without any vehicle or any other information enveloping component. However, here too the same basic model of mobility cycles applies, consisting of IT, origins, routes, controls, and destinations. Therefore, it will also be possible to assess future mobility modes and technologies in light of this same model, and thus making it possible to compare future mobility technologies to current and past mobility modes and technologies.

A wide array of terrestrial mobility media are available for the daily mobilities of contemporary mobility agents, notably in developed economies, whether public (trolleys; light trains; metros; trains; and buses of various sizes) or personal (walking; motorized and non-motorized cycling, and cars of various sizes). It turns out that whenever it is possible economically for mobility agents to purchase and maintain cars, this is the preferred medium for corporeal mobility, even when commuting is performed via public transportation. The car as an individual, comfortable and personally dominated and sheltered, mobility medium is highly attractive in contemporary developed society. Cycling is a preferred mode mainly in specific cities or cultures.

The tremendous development of personal virtual mobility media is yet another dimension of daily spatial mobilities of a personal nature, as part and parcel of a society dominated by individualism. The telephone, the Internet and the mobile phone are the three currently available technologies for speedy oral and written communications continuously available to their adopters for daily mobilities, with the mobile phone being the most widely adopted one worldwide. The smartphone integrates the mobile phone with the Internet, thus providing location-free access to both communications and information.

'Cyberspace', as the platform of the Internet, constitutes a rather complex entity, serving both the information and communications components of the Internet. The double nature of the Internet has made the experiencing and the cognition of cyberspace become more complex, involving multiple virtual experiences through websites, or information space, side-by-side with rather diversified interactions with communications partners, through electronic communications cyberspace. Instant access to cyberspace from anywhere by broadband users enhances the integration of cyberspace with real space from the perspective of Internet users. The instant connectivity to the Internet via broadband services implies that the 'terminals', in the form of two or more communicating parties or an Internet user and the websites he/she use, are now at stake rather than the 'roads' or channels connecting them.

Numerous cutting-edge technologies, as well as the technological innovations still being developed towards future adoption are aimed at mobile broadband uses. For example, remote access to one's desktop computer, notably to the 'my documents' file; the development of folding screens; and the development of Tablet computers, equipped with extremely high-resolution screens, encouraging electronic book reading on the go. These and other innovations and applications will encourage wider adoption of mobile Internet devices, culminating in an inability to lead personal and household lives without a location-free use of mobile Internet devices.

Individual business travel, as a sub-class of business travel, constitutes a routine mobility activity for numerous workers, involving the use of planes, hotels and other facilities and services which simultaneously serve also leisure tourism. Clear-cut differentiations between business and leisure tourisms have blurred for all the three major dimensions of tourism: people; places and activities. For people, business meetings by business people may yield leisure visits by these business persons and *vice versa*, leisure visits may bring about business ideas and opportunities yielding future business visits by vacationers. As for places and activities, leisure tourists and business visitors may share the same transportation, lodging and entertainment facilities. This blurring of boundaries between business and pleasure tourisms is similar to the emerging blurring of distinctions between daily home and work activities, so that mobility at large, domestically and internationally, routine and non-routine alike, evolves as a rather continuous and permanent state of life bringing about emerging integrations between business and leisure (see Urry 2000, Kellerman 2006a). The routine interface between business and touristic mobilities applies, though, only to a small segment of daily mobility agents: those active in routine business travel.

The universal adoption of mobile communications devices and their routine use for daily mobilities may have several implications regarding urban spatial reorganization. First, human beings are not ready, and maybe will never be ready, for a potential dismantling of cities when ICTs are ubiquitously adopted and services potentially acquired virtually. Second, mobile communications media do not present any extensive spatial requirements for their very functioning, so that they do not constitute a significant urban land-use. Third, mobile communications media are indifferent to existing urban structures and spatial organization of any kind in that they can function under any existing or future urban spatial structures. Fourth, the media for corporeal mobility in cities have not changed side by side with the introduction of ICTs. Fifth, city dwellers and visitors may move about any urban space in smarter and more efficient ways using GPS and mobile phones. Sixth, urbanites are in process of change of habits due to the universal use of mobile communications media, in that people's time can be used more flexibly. By the same token, technology has come to be used constantly by urbanites.

Transportation terminals turn out to be a must for all forms of public transportation used for daily corporeal mobilities, whether terrestrial or aerial. It further turns out that the impact of transportation terminals on urban life is a double one, in their constitution as nodes and as places. In their role as places, terminals attract commercial and business activity. Despite their location outside of cities, airport terminals, notably international ones, have a special importance for contemporary mobility, which is growingly based also on aerial travel. Airports tend to function under strict controls, notably security control, which heavily influences, and frequently dictates the physical structure of terminal environments and their operations, as well as passengers' flows and socialities, thus serving as a special spatial entity of human activity.

Widely adopted fixed and mobile Internet-based communications devices may have changed the long prevailing perception and status of space and distance as dictating a locational tyranny or destiny. Locations for daily activities which were limited geographically by convenient physical reach, have turned into locational opportunity, in the sense that many activities can now be performed at any distance with the use of the Internet. Flexible locational opportunities for individuals have become available, for instance, for online shopping, e-learning, home-based business/work, and social networking. However, the very use of these new and geographically flexible locational opportunities has only partially, though growingly, been adopted, except for the massive adoption of virtual social networking.

This tendency reflects a hesitation by individuals to change long established habits for local and domestic geographical anchoring of basic human activities. There still exist, thus, a gap between potential virtual daily mobilities and their materializations. The balance between local daily activities of individuals and geographically extended ones can potentially change drastically through a more extensive use of the Internet. It turns out that the very availability of virtual opportunities for daily mobilities, or technology-based locational flexibility, does not yet imply their use and adoption, even if the same infrastructures of Internet connectivity are widely adopted for flight and hotel reservations or for banking activities.

The daily, the spatial and the mobile

In this section we will attempt to present concluding thoughts and some ideas for future delving into daily spatial mobilities. We will do so through separate discussions of each of the three major components of daily spatial mobilities: the daily; the spatial and the mobile.

Daily and non-daily spatial mobilities

The discussions in the previous chapters call for some brief comparison between daily and non-daily mobilities, and the topic deserves separate rather thorough treatment. Daily spatial mobilities and non-daily ones, notably tourism and migration, share the same infrastructures of mobility media. Thus, the same information and communications systems consisting of Internet websites, e-mailing, SMS and voice calling are used for daily mobilities *per se*, as well as for preparations for physical tourism and migration movements. Similarly, the same media used for corporeal daily mobilities, mainly cars, trains and planes, are used for non-daily ones as well. However, the share of air travel is much higher for non-daily mobility than for daily, given the international destinations for most of the movement of tourists and migrants. Illegal migrants tend to walk or use boats which might permit crossing borders with weaker controls.

Globalization has been considered a major factor in the contemporary emergence of the massive movements of tourists and migrants, similarly to daily spatial mobilities (see e.g. Cresswell 2006a, Urry 2007). Also in similarity with daily mobilities, the development and mass adoption of virtual communications and information media may provide locational opportunities for non-daily mobilities as well. Tourism may be fostered by the extensive information provided on touristic sites on the Web, coupled with the ability to make reservations and produce tickets and boarding passes for flights and vouchers for ground services on the Web or over the telephone. Migrants too can find much information on potential destinations, as well as receive informal information when corresponding with friends and relatives who have already migrated.

There are, though, some major differences between daily and non-daily mobilities, which deserve separate detailed treatment, and will only be briefly highlighted here. The 'push' and 'pull' effects for daily mobilities discussed in Chapter 2 apply to non-daily ones only partially or in different ways. For example, the need for locomotion expressed in daily mobilities might not apply to migration at all and only to some forms of touristic travel which involve some form of hiking. On the other hand, the search for proximity may characterize numerous migrants wishing to join friends and relatives who migrated before them to a specific location.

Potential non-daily mobilities might also be different than potential daily ones, since the competences and appropriation dimensions of non-daily mobilities are of a different nature than those pertinent for daily mobilities, yielding sets of potential destinations for touristic travel and migration on an individual basis. Generally speaking, non-daily movements are much 'heavier' than the daily ones. Touristic travel and migration require preparations, frequently quite extensive ones. Preparations for daily mobilities might be non-existent at all or kind of 'automatically' performed by individuals when almost on the move. Linguistic competence might too be of special significance for non-daily mobilities. Following travel, non-daily travel may constitute a strong experience for travelers, and the memories from such travels are deeper and more significant than those of daily trips.

Non-daily mobilities may strongly affect the spatial reorganization of cities as compared to daily ones notably the virtual ones. Tourism may bring about the emergence of hotel and entertainment sections of cities, as well as the preservation of historical sites and the development of contemporary touristic attractions. Heavy migration into cities may bring about the development of segregated migrant neighborhoods or sections of a city, notably poorer ones, marked by landmarks, language and atmosphere per migrants' cultures.

Daily mobilities and space

Malecki and Moriset (2008) refer to the contemporary space, or world of places, as 'slippery' in that distance lost much of its significance, potentially bringing about both agglomeration and dispersion simultaneously (pp. 174-5). By the same token, contemporary daily spatial mobilities seem to have brought about an enhanced and upgraded space of flows turning it into a possible space of the mobile [people]. As we have saw in Chapter 1 the space of flows was defined as "the material organization of time-sharing social practices that work through flows" (Castells 2000: 442), in which 'flows' include all possible ones, except for people: capital, information, technology, organizational interaction, images, sounds, and symbols. However, the space of flows includes a certain segment of people in its third layer. The space of flows, as developed by Castells (2000: 448), consists of three layers: the first is technology, constituting a *circuit of electronic exchanges* embodied in networked cities; the second is a layer of places, *nodes and hubs*, hierarchically organized and topped by global cities, which serve as major loci of information production, and the third layer is people, the *managerial elites*, charged with the directional functions for the space of flows. Thus, in his proposed space of flows, Castells (2000) emphasized the 'material organization' as the core of the space of flows (consisting of communications infrastructures and cities), with only a small number of top managers directing the flows.

The current pattern and organization of flows, has emerged differently in that the space of flows includes most people, not only the managerial elite, operating daily, through e-mail, the Web, SMS, and phone calls. Furthermore, these flows of information do not concentrate anymore in specific nodes, mainly offices, but they rather take place everywhere in space, mainly within urban ones, including people on the move physically. The role of flow managers is less significant when all the communications systems are shared by a wide spectrum of the population. Contemporary information space consists of the flows of information side by side with the numerous mobile people who transmit and receive this information placelessly. Thus, on top of the communications and spatial infrastructures there seems to emerge the *space of the mobile*, of people who are constantly on the move, if not corporeally than virtually, or possibly both. As we have noted, the urban space of the mobile people needs no specific reorganization as in previous waves of urban development. However, this space of the mobile people is being extended and distanciated for individuals per their individual communications patterns.

The other side of the coin of the space of mobile people is mobile people in space. These mobile people enjoy much personal autonomy if they use personal mobility media such as cars and mobile phones, and if they prefer or have to use public transportation then they may make use of transportation terminals which house numerous commercial and public facilities and services in them. In any case, the speed of movement is a key element for contemporary mobile agents whether moving corporeally or virtually.

The mobile agent

The mobile agent *per se* has been portrayed and interpreted extensively elsewhere (e.g. Urry 2007, Cresswell 2006a, Kaufmann 2002). The discussion here will aim, therefore, specifically at highlighting some specific mobile agents, namely the daily and spatially mobile agents, jointly with daily spatial mobility as such. Obviously, individuals constitute, simultaneously, spatial as well as non-spatial (mainly social) mobile agents, and many of them constitute, once in a while, also non-daily but spatial mobile agents (as tourists), whereas others might at certain points in their lifecycle be migrants. However, given the continuous occurrence of daily spatial mobilities, and the safe assumption that most individuals in contemporary society are daily spatial mobile agents, it is of significance to take a separate look at these agents.

We may interpret the discussions in the previous chapters as having presented daily spatially mobile agents as involved in a process of social structuration. Structuration theory was proposed by Giddens (1979, 1981). At its core the theory suggests that social structures (as rules, norms and values) and infrastructures (e.g. religion, culture) on the one hand, and human agency on the other, operate continuously, simultaneously and interactively in society. In other words, daily human agency at large occurs and is practiced under the impact of socioeconomic structures and infrastructures (or superstructures), while at the same time, the cumulative actions of individuals create new structures. Thus, structure presents a duality by which it constitutes simultaneously both a medium and an outcome of social activity. This duality was termed *the duality of structure* (Giddens 1979: 5, 1981: 19). Gregory's (1978: 88-9) definition for structuration puts it directly within human spatial action: "Man is obliged to appropriate his material universe in order to survive, and [because] he is himself changed through changing the world around him in a continual and reciprocal process".

By the very nature of structuration theory, the ongoing structuration process for daily spatial mobility ties together the relevant mobility agents on the one hand, with the very social entity of daily spatial mobility on the other. The superstructures which pertain to mobile agents are mainly culture and religion which we discussed regarding personal mobility and personal autonomy (Chapter 3) and regarding potential mobilities (Chapter 4). Culture and religion may restrict one's autonomy and, thus, one's potential and actual levels and modes of mobility, or one's actual performance as a mobile agent.

The most basic structure for daily spatial mobility is the contemporary societal expectation from individuals to be constantly on the move, whether physically, virtually, or both. This expectation is expressed, at its basic level, through the emergence of proper mobility infrastructures: individuals purchase personal mobility media, and society, through governments and commercial companies, provide a universal availability of connectivity. The adoption of mobility media and the ability to use them anywhere puts humans increasingly in dynamic conditions of co-presence.

At yet another level, a society being constantly on the move has involved the emergence of three leading elements as far as the dimensions of the expected constant mobility by individuals are concerned. First is the social value of moving increasingly faster, so that higher mobility speeds imply more intensive movements. Second is globalization which points to the expected spatial extent of movements, stretching to its widest possible extent, and finally, the third dimension of continuous mobility is social networking, which implies an expected wide social spectrum of connectivity.

Each daily mobile agent acts within this structural system in what seems to be an individual pattern of mobility, changing from one person to another regarding the pace of one's mobility and its spatial and social spreads or extents. However, it seems also that all these three parameters have been transformed for all daily and spatially mobile agents through the influence of social structures (speed, globalization and networking), notably regarding virtual mobility. For example, the very location of websites visited by individuals is usually unknown and not even of interest to their users, so that users are led through global paths automatically. The extremely fast growth of global social networks, side by side with the growing globalization of business and work, make the seemingly individual habits of mobility fit structural lines. On the other hand, the aggregated mobility habits of individuals strengthen and amplify the structural accent on constant mobility as expressed in increased speeds of connectivity, globalization, and networking. The mobility industry provides ever faster media and power coupled with tools and frameworks for more sophisticated and globally stretched networking. This is the duality of structure as far as daily spatial mobility and daily spatial mobile agents are concerned.

Management of daily spatial mobilities

At first glance it seems that the very performance of daily spatial mobilities by an endless number of individuals, each having their own mobility patterns and habits, refutes a possible management of daily spatial mobilities. However, the need for some management of these mobilities is of great importance for both individuals and society, given the diversity and complexity of contemporary daily mobilities, both corporeal and virtual spatial.

For individuals in developed countries there are several contemporary mobility trends which call for personal management of mobility: Three of these dominate the field. First is the growing adoption of smartphones which continuously become smarter information and communications machines and imply constant availability of information and communications. Second is the almost ubiquitous involvement in social networking which implies a heavy investment of time. Third is the personal availability of private cars which provide more flexibility for corporeal mobility.

Individuals have, thus, to cope with several decision-making tasks which are either completely novel, or have gained more central roles in recent daily lives. Examples of such tasks are: which matters should be performed physically and which ones should be executed virtually? If the virtual path is preferred, then which medium should be chosen for any matter? How to coordinate between private communications, on the one hand, and work/study and family responsibilities, on the other? When and how to move from virtual social communications to face-to-face? It is expected, therefore, that in the near future some management ideas will develop either from the bottom-up, namely by cumulative human agency, or by management theory/expertise or public rules, or possibly by both, implying a new round of the continuous structuration process. By the same token, children are exposed to communications and information media almost from infancy. Proper use and management of information, as well as proper conduct of communications, will have to become a skill taught in schools.

From a public perspective, governments and major private companies maintain the daily spatial mobilities of people at the metropolitan, national and global levels in sophisticated ways unknown in the past, using complex computing systems, radar, cameras, etc. However, mobility disruptions still occur. Traffic jams in major urban arterial roads have become a daily experience which urbanites have to take into account when using their cars, notably during rush hours. However, the collapse of a mobile phone system, as happens once in a while, can be almost disastrous for huge numbers of subscribers as well as for those calling these subscribers. By the same token, major stopping of air traffic, such as caused by the volcanic ash from Iceland in 2010, can have widespread repercussions for air travel in general, both internal and inter-continental.

Another type of mobility disruption, notably under conditions of advanced globalization, can be global economic crises, such as during 2008–9 (Kellerman 2009b). Such a crisis may bring about changes in daily spatial mobilities of individuals with extensive repercussions for mobility business. The recent crisis did not hit equally all the branches of the mobility economy. It has turned out that times of crisis imply a much stronger need by numerous individuals for updated, on-time business and financial information in order to cope quickly with changing circumstances. This is so for businesses as well as for households, mainly in the developed world. The instant availability of swiftly changing financial information is sometimes of critical value, given the growing fluctuations and nervousness of the financial markets as compared to past crises, since all markets worldwide are exposed on-time and on-line with full transparency to Internet users whether connected through PCs, laptops or mobile phones. This on-time connectivity permits instant reactions, especially in stock markets and in foreign exchange ones.

Thus, ISPs, *Google* and mobile telephone companies presented revenues even during the crisis, as compared to other branches of the mobility economy dealing with corporeal mobility. At the same time, car manufacturers in both North America and Europe got either close to bankruptcy or reported significantly lowered sales, which was also true for airline companies. *Google* is a special case in this regard as it provides its services free of charge to users, while earning its income from advertisements, which might decline with the decrease in economic activity. Under severe economic circumstances which do not permit generous travel budgets for senior staff, video conferences and videophone calls can serve as substitutes for face-to-face meetings, notably for in-company discussions. Despite the loss of informal atmosphere, body language and gestures, the use of video transmissions and even the use of more indirect written e-mail messages might be called for. At least, the very availability of well-developed visual virtual mobility media may permit companies to consider corporeal versus virtual meetings at a time of economic crisis. On a more restricted scale, video calls may serve also as a substitute for travel for social purposes as well.

Another dimension of the continuous flourishing of virtual mobility at times of crisis was the introduction of new products and technologies for virtual mobility which continued even during the crisis. It was at the February 2009 Barcelona Mobile World Congress and fair that major global producers of mobile phones introduced expensive touch-screen models, permitting more comfortable work with the Internet. One could think that the global crisis would have made customers spend less money on the purchase of more sophisticated devices and services. This continued introduction of sophisticated and rather expensive mobile phones may reflect, at least partially, a response to demand for highly mobile Internet services permitting instant use anywhere. A similar trend was the introduction of VoIP software for 3rd generation mobile phones, which include connection to the Internet. These developments permitted cheaper transmission of business information to and by individuals and they might have saved some corporeal travel.

Another level of potential mobility management is the, theoretical at least, management of mobility innovations. Technology-based mobilities seem to undergo endless advancement, either for more environmentally friendly automobiles or for more sophisticated mobility ICTs. However, any surveys which are performed before the development of a technology for mass production cannot fully predict the actual adoption of new mobility media or technological modifications in them. For example, videophones were originally introduced in 1964, and have been reintroduced time and again since then, but the use of video for telephone conversations only became widely adopted in the late 2000s as a component of the more recent introduction of free, or modestly priced, VoIP on smartphones, coupled with free Wi-Fi connectivity for mobile phones. These innovations, on their part, were based on the earlier mass adoption of both mobile phones (a device) and social networking (a societal trend). It is, therefore, difficult to manage *a priori* the possible wide adoption of new technologies even if a 'need' for their use is identified by entrepreneurs.

References

Aaltola, M. 2005. The international airport: The hub-and-spoke pedagogy of the American empire. *Global Networks*, 5, 261-78.

Achille, S.J. 2008. World statistics on the number of Internet shoppers. http://www.multilingual-search.com/world-statistics-on-the-number-of-internet-shoppers/28/01/2008/.

ACI (Automobil Club d'Italia). 2005. Personal communications with A.Vasserot.

Adams, P.C. 1995. A reconsideration of personal boundaries in space-time. *Annals of the Association of American Geographers*, 85, 267-85.

Adams, P.C. 2001. Peripatetic imagery and peripheral sense of place, in *Textures of Place: Exploring Humanist Geographies*, edited by P.C. Adams, S. Hoelscher, and K. Till. Minneapolis: University of Minnesota Press.

Adams, P.C. 2009. *Geographies of Media and Communication: A Critical Introduction*. Chichester: Wiley-Blackwell.

Adams, P.C. and Ghose, R. 2003. India.com: The construction of a space between. *Progress in Human Geography*, 27, 414-37.

Adey, P. 2004a. Surveillance at the airport: Surveilling mobility/mobilising surveillance. *Environment and Planning A*, 36, 1365-80.

Adey, P. 2004b. Secured and sorted mobilities: Examples from the airport. *Surveillance and Society*, 1, 500-19.

Adey, P. 2006a. Airports and air-mindedness: Spacing, timing and using the Liverpool Airport, 1929-1939. *Social and Cultural Geography*, 7, 343-63.

Adey, P. 2006b. If mobility is everything then it is nothing: Towards a relational politics of (im)mobilities. *Mobilities*, 1, 75-94.

Adey, P. 2007. 'May I have your attention': Airport geographies of spectatorship, position, and (im)mobility. *Environment and Planning D: Society and Space*, 25, 515-36.

Adey, P. 2010. *Mobility*. London and New York: Routledge.

Agar, J. 2003. *Constant Touch: A Global History of the Mobile Phone*. Cambridge: Revolutions in Science.

Allen, J. 1999. Worlds within cities, in *City Worlds*, edited by D. Massey, J. Allen, and S. Pile. London: Routledge, 53-98.

Allot Communications. 2010. Allot mobile trends: Global mobile broadband traffic report H2, 2009. http://www.allot.com.

Altman, I. 1975. *The Environment and Social Behavior: Privacy, Personal Space, Territory, Crowding*. Monterey, CA: Brooks/Cole.

Amdur, L. and Epstein-Pliouchtch, M. 2009. Architects' places, users' places: Place meanings at the new central bus station, Tel Aviv. *Journal of Urban Design*, 14, 147-61.

American Public Transportation Association. 2010. Public Transportation Fact Book. http://www.apta.com.

Amin, A. and Thrift, N. 2002. *Cities: Reimagining the Urban*. Cambridge: Polity.

Appadurai, A. 1990. Disjuncture and difference in the global cultural economy. *Theory, Culture and Society*, 7, 295-310.

Aoyama, Y. 2003. Sociospatial dimensions of technology adoption: Recent M-commerce and E-commerce developments. *Environment and Planning A*, 35, 1201-21.

Arminen, I. 2007. Review essay: Mobile communication society? *Acta Sociologica*, 50, 431-7.

Augé, M. 2000. *Non-Places: Introduction to an Anthropology of Supermodernity*, translated by J. Howe. London: Verso.

Aurnague, M., Hickmann, M. and Vieu, L. 2007. Introduction: Searching for the categorization of spatial entities in language and cognition, in *The Categorization of Spatial Entities in Language and Cognition*, edited by M. Aurnague, M. Hickmann and L. Vieu. Philadelphia: John Benjamin Publishing Company, 1-32.

Avidan, I. and Kellerman, A. 2004. Distance in the Internet by time and route: An empirical examination. *Contemporary Israeli Geography. Horizons* 60-1: 77-88.

Bachmair, B. 1991. From the motor-car to television: Cultural-historical arguments on the meaning of mobility for communication. *Media, Culture and Society*, 13, 521-33.

Bán, D. 2007. The railway station in the social sciences. *The Journal of Transport History*, 28, 289-93.

Banister, D. 2011. The trilogy of distance, speed and time. *Journal of Transport Geography*, 19, 950-9.

Barnett, M.W. and Larkman, P.M. 2007. The action potential. *Practical Neurology*, 7, 192-7.

Barrett, C. 2005. Keynote address at the second phase of WSIS (World Summit on the Information Society), Tunis, 16 November http://www.itu.int/WSIS/tunis/statements/docs/ps-intel-opening/1.pdf.

Bates, P. and Huws, U. 2002. Modelling ework in Europe: Estimates, Models and Forecasts from the EMERGENCE Project. Falmer, Brighton: Institute for Employment Studies. http://www.employment-studies.co.uk/pdflibrary/388.pdf.

Batty, M. 1997. Virtual geography. *Futures*, 29, 337-52.

Baudrillard, J. 1988. *Jean Baudrillard: Selected Writings*, edited by M. Poster. Cambridge and Stanford: Polity and Stanford University Press.

Baudrillard, J. 1996. *The System of Objects*. London: Verso.

Bauman, Z. 2000. *Liquid Modernity*. Cambridge: Polity Press.

Beaverstock, J.V., Derudder, B., Faulconbridge, J. and Witlox, F. (eds.). 2010. *International Business Travel in the Global Economy*. Farnham: Ashgate.

Beaverstock, J.V., Derudder, B., Faulconbridge, J. and Witlox, F. 2010. International business travel and the global economy: Setting the context, in *International*

Business Travel in the Global Economy, edited by J.V. Beaverstock, B. Derudder, J. Faulconbridge, and F. Witlox. Farnham: Ashgate, 1-7.

Beck, U. and Beck-Gernsheim, E. 2001. *Individualization: Institutionalized Individualism and its Social and Political Consequences*. London: Sage.

Beckmann, J. 2001. Automobility – a social problem and theoretical concept. *Environment and Planning D: Society and Space*, 19, 593-607.

Benedikt, M. 1991. Cyberspace: Some proposals, in *Cyberspace: First Steps*, edited by M. Benedikt. Cambridge, MA: MIT Press, 119-224.

Ben-Ze'ev, A. 2004. *Love Online: Emotions on the Internet*. Cambridge: Cambridge University Press.

Berlin, I. 1969. *Four Essays on Liberty*. Oxford: Oxford University Press.

Bertolini, L. 1996. Nodes and places: Complexities of railway station redevelopment. *European Planning Studies*, 4, 331-45.

Bertolini, L. 1999. Spatial development patterns and public transport: The application of an analytical model in the Netherlands. *Planning, Practice & Research*, 14, 199-210.

Bertolini, L. and Spit, T. 1998. *Cities on Rails: The Redevelopment of Railway Station Areas*. London: E& FN Spon.

Bertolini, L. and le Clercq, F. 2003. Urban development without more mobility by car? Lessons from Amsterdam, a multimodal urban region. *Environment and Planning A*, 35, 575-89.

Biology-Online. 2010. Motility. http://www.biology-online.org/dictionary/Motility.

Biology-Online. 2010. Sessile. http://www.biology-online.org/dictionary/Sessile.

Bissell, D. 2009. Conceptualizing differently-mobile passengers: Geographies of everyday encumbrance in the railway station. *Social & Cultural Geography*, 10, 173-95.

Bissell, D., Adey, P. and Laurier, E. 2011. Introduction to the special issue on geographies of the passenger. *Journal of Transport Geography*, 19, 1007-9.

Blainey, A. 1966. *Tyranny of Distance: How Distance Shaped Australia's History*. Melbourne: Macmillan.

The Blog Herald. 2005. http://www.blogherald.com/2005/10/10/the-blog-herald-blog-count-october-2005/.

Blomley, N.K. 1994. *Law, Space, and the Geographies of Power*. New York: Guilford.

Blumen, O. and Kellerman, A. 1990. Gender Differences in Commuting Distance, Residence and Employment Location: Metropolitan Haifa 19721983. *The Professional Geographer*, 42, 5471.

Boden, D. and Molotch, H.L. 1994. The compulsion of proximity, in *NowHere Space, Time and Modernity*, edited by R. Friedland and D. Boden. Berekeley: University of California Press, 257-86.

Böhm, S., Jones, C., Land, C., and Paterson, M. 2006. Introduction: Impossibilites of automobility. *Sociological Review*, 54, 3-16.

Boltanski, L. and Chiapello, È. 2007. *The New Spirit of Capitalism*. Translated by G. Elliot. London and New York: Verso.

Bolter, J.D. and Grusin, R. 1999. *Remediation: Understanding New Media.* Cambridge: MIT Press.

Bonss, W. 2004. Introduction. The mobility and the cosmopolitan perspective workshop. Munich Reflexive Modernization Research Centre.

Bonss, W. and Kesselring, S. 2001. Mobilität am Übergang von der Ersten zur Zweiten Moderne, in *Die Modernisierung der Moderne*, edited by U. Beck and W. Bonss. Frankfurt am Main: Suhrkamp, 177-90.

Bonss, W. and Kesselring, S. 2004. *Mobility and the cosmopolitan perspective.* Paper to the mobility and the cosmopolitan perspective workshop, Munich, Reflexive Modernization Research Centre.

Bourdieu, P. 1984. *Distinction: A Social Critique of the Judgment of Taste.* London: Routledge and Kegan Paul.

Bourdieu, P. 1987. What makes a social class? On the theoretical and practical existence of groups. *Berkeley Journal of Sociology*, 22, 1-17.

Braidotti, R. 1994. *Nomadic Subjects: Embodiment and Sexual Difference in Contemporary Feminist Theory.* New York: Columbia University Press.

Brandon, R. 2002. *Auto Mobile.* London: Macmillan.

Breton, G. and Lambert, M. (eds). 2003. *Universities and Globalization: Private Linkages, Public Trust.* Paris: UNESCO and Université Laval.

Buliung, R.N. 2011. Wired people in wired places: Stories about machines and the geography of activity. *Annals of the Association of American Geographers* 101: 1365-81.

Burnett, P. and Lucas, S. 2010. Talking, walking, riding and driving: The mobilities of older adults. *Journal of Transport Geography*, 18, 596-602.

Business Insider. 2011. Facebook has more than 600 million users, Goldman tells clients. http://www.businessinsider.com/facebook-has-more-than-600-million-users-goldman-tells-clients-2011-1.

Cairncross, F. 1997. *The Death of Distance: How the Communications Revolution Will Change Our Lives.* Boston: Harvard Business School Press.

Canzler, W. 2004. The vision of mobility and second modernity. Paper presented at the mobility and the cosmopolitan perspective workshop. Munich Reflexive Modernization Research Centre.

Canzler, W., Kaufmann, V. and Kesselring, S. 2008. Tracing mobilities: An introduction, in *Tracing Mobilities: Towards a Cosmopolitan Perspective*, edited by W. Canzler, V. Kaufmann, and S. Kesselring. Aldershot: Ashgate, 1-12.

Cass, N., Shove, E. and Urry, J. 2005. Social exclusion, mobility and access. *The Sociological Review*, 53, 539-55.

Castells, M. 1989. *The Informational City: Information, Technology, Economic Restructuring and the Urban-Regional Process.* Oxford: Blackwell.

Castells, M. 1996. *The Rise of the Network Society, Volume 1: The Information Age: Economy, Society and Culture.* Oxford: Blackwell.

Castells, M. 1998. *End of Millennium.* Oxford: Blackwell.

Castells, M. 2000. *The Rise of the Network Society.* 2nd ed. Oxford: Blackwell.

Castells, M. 2001. *The Internet Galaxy: Reflections on the Internet, Business, and Society*. New York: Oxford University Press.

Castells, M. 2009. *Communication Power*. Oxford: Oxford University Press.

Castells, M., Fernánddez-Ardèvol, M., Qiu, J.L., and Sey, A. 2007. *Mobile Communication and Society: A Global Perspective*. Cambridge, MA: MIT Press.

CEOS for Cities. 2010. New York's green dividend. http://www.ceosforcities.org/work/nycs_green_dividend.

Chang, Y-L. 2003. Spatial cognition in digital cities. *International Journal of Architectural Computing*, 1, 471-88.

Cidell, J. 2006. Air transportation, airports, and the discourses and practices of globalization. *Urban Geography*, 27, 651-63.

Citymayors Statistics. 2008. http://www.citymayors.com/statistics/urban_2006_1.html.

City of Westminster. 2006. http://www.westminster.gov.uk/leisureandculture/tourismandtravel/businessvisitors.cfm.

Cohen, E. 1974. Who is a tourist? A conceptual clarification. *Sociology*, 22, 527-55.

Comer, J.C. and Wikle, T.A. 2008. Worldwide diffusion of the cellular telephone, 1995-2005. *The Professional Geographer*, 60, 252-69.

Commission of the European Communities. 2009. Report on cross-border e-commerce in the EU. http://ec.europa.eu/consumers/strategy/docs/com_staff_wp2009_en.pdf.

Cooper, G. 2001. The mutable mobile: Social theory in the wireless world, in *Wireless World: Social and Interactional Aspects of the Mobile Age*, edited by B. Brown, N. Green, and R. Harper. London: Springer, 19-31.

Couclelis, H. 1998. Worlds of Information: The geographic metaphor in the visualization of complex information. *Cartography and Geographic Information Systems*, 25, 209-20.

Couclelis, H. 2004. Pizza over the Internet: E-commerce, the fragmentation of activity and the tyranny of the region. *Entrepreneurship and Regional Development*, 16, 41-54.

Couclelis, H. 2009. Rethinking Time Geography in the Information Age. *Environment and Planning A*, 41, 1556-75.

Crang, M. 2002. Between places: Producing hubs, flows, and networks. *Environment and Planning A*, 34, 569-74.

Crang, M., Crang, P., and May, J. (eds). 1999. *Virtual Geographies: Bodies, Space and Relations*. London: Routledge.

Cresswell, T. 2001a. The production of mobilities. *New Formations*, 43, 11-25.

Cresswell, T. 2001b. *The Tramp in America*. London: Reaktion.

Cresswell, T. 2002. Introduction: Theorizing place, in *Mobilizing Place, Placing Mobility: The Politics of Representation in a Globalized World*, edited by G. Verstraete and T. Cresswell. Amsterdam: Rodopi, 11-32.

Cresswell, T. 2004. *Place: A Short Introduction*. Oxford: Blackwell.

Cresswell, T. 2006a. *On the Move: Mobility in the Modern Western World.* New York and London: Routledge.

Cresswell, T. 2006b. The right to mobility: The production of mobility in the courtroom. *Antipode*, 38, 735-54.

Crutcher, M. and Zook, M. 2009. Placemarks and waterlines: Racialized cyberscapes in post-Katrina Google Earth. *Geoforum*, 40, 523-34.

Cwerner, S. 2009. Introducing aeromobilities, in *Aeromobilities*, edited by S. Cwerner, S. Kesselring and J. Urry. London and New York: Routledge, 1-22.

Dant, T. and Martin, P.J. 2001. By car: Carrying modern society, in *Ordinary Consumption*, edited by J. Gronov and A. Warde. London: Routledge.

Davidson, R. 2000. *Business Travel*. Essex: Longman/Pearson Education.

Davidson, R. and Cope, B. 2003. *Business Travel: conferences, Incentive Travel, Exhibitions, Corporate Hospitality and Corporate Travel.* Harlow: Prentice Hall.

Davis, G.L.H. 1989. On the nature of geo-history, with reflections on the historiography of geomorphology, in *History of geomorphology*, Binghampton Symposium No. 19, edited by K.J. Tinkler. Boston: Unwin, 103-9.

De Certeau, M. 1985. Practices of space, in *On Signs*, edited by M. Blonsky. Oxford: Basil Blackwell.

De Gournay, C. 2002. Pretense of intimacy in France, in *Perpetual contact: Mobile communication, private talk, public performance*, edited by J. Katz and M. Aakhus, Cambridge, MA: MIT Press, 193-205.

Demarco, C.W. 2004. The generation and destruction of categories, in *Categories: Historical and Systematic Essays*, edited by M. Gorman and J.S. Sanford. Washington, DC: The Catholic University of America Press, 238-67.

Dicken, P. 2007. *Global Shift: Mapping the Changing Contours of the World Economy*. 5th ed. New York: Guilford.

Dodge, M. 2001. Guest editorial. *Environment and Planning B: Planning and Design*, 28, 1-2.

Dodge, M. and Kitchin, R. 2001. *Mapping Cyberspace*. London: Routledge.

Dodge, M. and Kitchin, R. 2004. Flying through code/space: The real virtuality of air travel. *Environment and Planning A*, 36, 195-211.

Dodge, M., Kitchin, R. and Zook, M. 2009. How Does Software Make Space? Exploring Some Geographical Dimensions of Pervasive Computing and Software Studies. *Environment and Planning A*, 41, 1283-93.

Drori, G.S. 2007. Information society as a global policy agenda: What does it tell us about the age of globalization? *International Journal of Comparative Sociology*, 48, 297-316.

Duranton, G. 1999. Distance, land, and proximity: Economic analysis and the evolution of cities. *Environment and Planning A*, 31, 2169-88.

eMarketer. 2011. Worldwide Internet users: 2005-2011. http://www.emarketer.com/Reports/All/Em_internet_feb07.aspx.

Engelbrekt, K. 2011. Mobility and the notion of attainable reach, in *The Politics of Proximity: Mobility and Immobility in Practice*, edited by G. Pellegrino. Farnham: Ashgate, 31-42.

European Commission. 2002. Qualitative study on cross border shopping in 28 European countries. http://ec.europa.eu/consumers/topics/cross_border_shopping_en_pdf.

European Commission. 2006. Consumer protection in the Internet market. Special Eurobarometer 252 http://ec.europa.eu/public_opinion/archives/ebs/ebs252_en.pdf.

European Commission. 2011. Consumer attitudes towards cross-border trade and consumer protection: Analytical report. Flash Eurobarometer 299. http://ec.europa.eu/public_opinion/flash/fl_299_wn.pdf.

EU (European Union). 2006. eUSER population survey 2005. http://www.euser-eu.org/eUSER_PopulationSurveyStatistics.asp.

Eurostat. 2008. Internet usage in 2008- households and individuals. http://epp.eurostat.ec.europa.eu/cache/ITY_OFFPUB/KS-QA-08-046/EN/KS-QA-08-046-EN.PDF.

Eyerman, R. and Löfgren, O. 1995. Romancing the road: Road movies and images of mobility. *Theory, Culture and Society*, 12, 53-79.

Fabrikant, S.I. and Buttenfield, B.P. 2001. Formalizing semantic spaces for information access. *Annals of the Association of American Geographers*, 91, 263-80.

Faulconbridge, J.R. and Beaverstock, J.V. 2007. Geographies of international business travel in the professional service economy. *GaWC Research Bulletin* 252. http://www.lboro.ac.uk/gawc/rb/rb252.html.

Featherstone, M. 2004. Automobiles: An introduction. *Theory, Culture and Society*, 21, 1-24.

FCC (Federal Communications Commission). 2009. Statement of Chairman Kevin J. Martin. http://hraunfoss.fcc.gov/edocs_public/attachmatch/DOC-280909A2.doc.

Felstead, A., Jewson, N., and Walters, S. 2005. *Changing Places of Work*. Basingstoke: Palgrave Macmillan.

Fischer, C.S. 1992. *America Calling: A Social History of the Telephone to 1940*. Berkeley: University of California Press.

Fischer, C.S. and Carroll, G.R. 1988. Telephone and automobile diffusion in the United States, 1902-1937. *The American Journal of Sociology* 93, 1153-78.

Flamm, M. and Kaufmann. V. 2006. Operationalising the concept of motility: A qualitative study. *Mobilities*, 1, 167-89.

Fortunati. L. 2002. Italy: Stereotypes, true and false, in *Perpetual Contact: Mobile Communication, Private Talk, Public Performance*, edited by J. E. Katz and M. A. Aakhus Cambridge: Cambridge University Press, 42-62.

Fowler, H. 1965. *Curiosity and Exploratory Behavior*. New York: Macmillan.

Francis, G., Fidato, A., and Humphreys, I. 2003. Airport-airline interaction: The impact of low-cost carriers on two European airports. *Journal of Air Transport Management*, 9, 267-73.

Francis, G., Humphreys, I., and Ison, S. 2004. Airports' perspectives on the growth of low-cost airlines and the remodeling of the airport-airline relationship. *Tourism Management*, 25, 507-14.

Freud, S. 1930. *Civilization and its Discontent*, translated by J. Riviere. London: Hogarth Press.

Freund, P. and Martin, G. 1993. *The Ecology of the Automobile*. Montreal: Black Rose Books.

Friedman, R. 1990. On the concept of authority in political philosophy, in *Authority*, edited by J. Raz. New York: New York University Press, 56-91.

Friedman, M. 2003. *Autonomy, Gender, Politics*. New York: Oxford University Press.

Fuller, G. and Harley, R. 2005. *Aviopolis: A Book about Airports*. London: Black Dog Publishing.

Gant, D. and Kiesler, S. 2002. Blurring the boundaries: Cell phones, mobility, and the line between work and personal life, in *Wireless World: Social and Interactional Aspects of the Mobile Age*, edited by B. Brown, N. Green, and R. Harper. London: Springer, 121-31.

Garreau, J. 1991. *Edge City*. New York: Doubleday.

Gartman, D. 2004. Three ages of the automobile. *Theory, Culture and Society*, 21, 169-95.

Gershuny, J. 2003. Web use and net nerds: A neofunctionalist analysis of the impact of information technology in the home. *Social Forces*, 82, 141-68.

Gessel, A. and Ilg, F. 1943. *Infant and Child in the Culture of Today*. New York: Harper and Row.

Gibson, W. 1984. *Neuromancer*. London: Gollancz.

Giddens, A. 1979. *Central Problems in Social Theory*. London: Macmillan.

Giddens, A. 1981. *A Contemporary Critique of Historical Materialism*. Berkeley: University of California Press.

Giddens, A. 1990. *The Consequences of Modernity*. Cambridge: Polity Press.

Giddens, A. 1992. *Modernity and Self-Identity: Self and Society in the Late Modern Age*. Cambridge: Polity Press, Cambridge.

Giordano, P.C. 2003. Relationships in Adolescence. *Annual Review of Sociology*, 29, 257-81.

Gkritza, K., Niemeier, D. and Mannering, F. 2006. Airport security screening and changing passenger satisfaction: An exploratory assessment. *Journal of Air Transport Management*, 12, 213-19.

Goffman, E. 1971. *Relations in Public: Microstudies of the Public Order*. Harmondsworth: Penguin.

Golledge, R.G. 1995. Primitives of spatial knowledge, in *Cognitive Aspects of Human-Computer Interaction for Geographic Information Systems*, edited by

T.L. Nyerges, D.M. Mark, R. Laurini, and M.J. Egehofer. Boston: Kluwer, 29-44.

Golledge, R.G. 1999. Human wayfinding and cognitive maps, in *Wayfinding Behavior: Cognitive Mapping and Other Spatial Processes*, edited by R.G. Golledge. Baltimore: Johns Hopkins University Press, 5-45.

Golledge, R.G. and Stimson, R.J. 1997. *Spatial Behavior: A Geographic Perspective*. New York: Guilford Press.

Gordon, A. 2004. *Naked Airport: A Cultural History of the World's Most Revolutionary Structure*. New York, Henry Holt.

Gordon, E. and de Souza e Silva, A. 2011. *Net Locality: Why Location Matters in a Networked World*. Chichester: Wiley-Blackwell.

Gorz, A. 1980. *Ecology as Politics*, translated by P. Vigderman and J. Cloud. Boston: South End Press.

Gottdiener, M. 2001. *Life in the Air: Surviving the New Culture of Air Travel*. Lanham, MD: Rowman and Littlefield.

GPS Magazine. 2009. http://www.gpsmagazine.com/2009/10/gps_us_household_penetration.php.

Graham, S. and Marvin, S. 1996. *Telecommunications and the City*. London and New York: Routledge.

Graham, S. and Marvin, S. 2001. *Splintering Urbanism: Networked Infrastructures, Technological Mobilities and the Urban Condition*. London: Routledge.

Greenfield, A. 2006. *Everware: The Dawning Age of Ubiquitous Computing*. Berkeley: New Riders Press.

Gregory, D. 1978. *Ideology, Science and Human Geography*. London: Hutchinson.

Gross, G. 2009. Study: US among world's leaders in broadband use. *PC World*, http://www.pcworld.com/printable/article/id,185417/printable.html.

Haddon, L. 2004. *Information and Communication Technologies in Everyday Life: A Concise Introduction and Research Guide*. Oxford: Berg.

Hägerstrand, T. 1970. What about people in regional science? *Papers of the Regional Science Association*, 24, 7-21.

Hägerstrand, T. 1992. Mobility and transportation – are economics and technology the only limits? *Facta and Futura*, 2, 35-8.

Halford, S. 2005. Hybrid workspace: Re-spatialisations of work, organisation and management. *New Technology, Work and Employment*, 20, 19-33.

Hannam, K., Sheller, M. and Urry, J. 2006. Editorial: Mobilities, immobilities and Moorings. *Mobilities*, 1, 1-22.

Hanson, S. 1995. Getting there: Urban transportation in context, in *The Geography of Urban Transportation*, edited by S. Hanson. New York: Guilford, 3-25.

Hanson, S. 1998. Off the road? Reflections on transportation geography in the information age. *Journal of Transport Geography*, 6, 241-9.

Harvey, D. 1989. *The Condition of Postmodernity*. Oxford: Blackwell.

Haynes, P. 2010. Information and communication technology and international business travel: Mobility allies? *Mobilities*, 5, 547-64.

Helminen, V. and Ristimäki, M. 2007. Relationships between commuting distance, frequency and telework in Finland. *Journal of Transport Geography*, 15, 331-42.

Henderson, J. 2004. The politics of mobility and business elites in Atlanta, Georgia. *Urban Geography*, 25, 193-216.

Hislop, D. and Axtell, C. 2007. The neglect of spatial mobility in contemporary studies of work: The case of telework. *New Technology, Work and Employment*, 22, 34-51.

Hochmair, H. and Frank, A.U. 2001. A semantic map as basis for the decision process in the www navigation, in *Conference on Spatial Information Theory*, edited by D.R. Montello. Morrow Bay, CA: Springer, 173-88.

Hubbard, P., Kitchin, R., Bartley, B., and Fuller, D. 2002. *Thinking Geographically: Space, Theory and Contemporary Human Geography*. London and New York: Continuum.

Hubbard, P. and Lilley, K. 2004. Pacemaking the modern city: The urban politics of speed and slowness. *Environment and Planning D: Society and Space*, 22, 273-94.

Hughes, T.P. 1987. The evolution of large technological systems, in *The Social Construction of Technological Systems: New Directions in the Sociology and History of Technology*, edited by W.E. Bijker, T.P. Hughes, and T.J. Pinch. Cambridge MA: MIT Press, 51-82.

Huh W-k. 2006. A geography of virtual universities in Korea. Paper presented at the Annual Meeting of the IGU Commission on the Geography of the Information Society, Sydney.

Hupkes, G. 1982. The law of constant travel time and trip-rates. *Futures*, 14, 38-46.

Hwang, W., Jung, H.-S, and Salvendy, G. 2006. Internationalisation of e-commerce: A comparison of online shopping preferences among Korean, Turkish and US populations. *Behaviour and Information Technology*, 25, 3-18.

Imrie, R. 2000. Disability and discourses of mobility and movement. *Environment and Planning A*, 32, 1641-56.

Ingold, T. 2004. Culture on the ground – the world perceived through the feet. *Journal of Material Culture*, 9, 315-40.

Ingold, T. and Vergunst J.L., editors. 2008. *Ways of Walking*. Aldershot: Ashgate.

IATA (International Air Transport Association). 2006. Scheduled passengers carried. http://www.iata.org/pressroom/wats_passengers_carried.htm.

IRF (International Road Federation). 2009. IRF World Road Statistics 2009. http://www.irfnet.org/files-upload/stats/2009/wrs2009_web.pdf.

ITU (International Telecommunication Union). 2003. The birth of broadband. http://www.itu.int/osg/spu/publications/birthofbroadband/faq.html.

ITU (International Telecommunication Union). 2010a. The world in 2009: ICT facts and figures. http://www.iyu.int/ITU-D/ict/material/Telecom09_flyer.pdf.

ITU (International Telecommunication Union). 2010b. Statistics. http://www.itu.int/ITU-D/ict/statistics/material/graphs/movile_reg-09.jpg.

ITU (International Telecommunication Union). 2011. The world in 2011: ICT facts and figures. http://www.itu.int/ITU-D/ict/facts/2011/material/ICTFactsFigures2011.pdf.

Internet World Stats. 2011. Internet usage statistics (2010). http://www.internetworldstats.com/stats1.htm.

Iyer, P. 2001. *The Global Soul: Jet Lag, Shopping Malls, and the Search for Home*. New York: Random House.

Jacobs, J. 1961. *The Death and Life of Great American Cities*. New York: Random House.

Janelle, D.G. 1968. Central place development in a time-space framework. *The Professional Geographer*, 20, 5-10.

Janelle, D.G. 1973. Measuring human extensibility in a shrinking world. *The Journal of Geography*, 72, 8-15.

Janelle, D.G. 1991. Global interdependence and its consequences, in *Collapsing Space and Time: Geographic Aspects of Communication and Information*, edited by S.D. Brunn and T.R. Leinbach. London: Harper Collins Academic, 49-81.

Janelle, D.G. 2004. Impact of information technologies, in *The Geography of Urban Transportation*, 3rd edition, edited by G. Hanson, and G. Giuliano. New York: Guilford, 86-112.

Janelle, G.J. 1969. Spatial reorganization: A model and concept. *Annals of the Association of American Geographers*, 59, 348-64.

JiWire. 2010. Global Wi-Fi finder. http://v4.jiwire.com/search-hotspot-locations.htm.

Jirón, P. 2011. Mobility practices in Santiago de Chile: The consequences of restricted urban accessibility, in *The Politics of Proximity: Mobility and Immobility in Practice*, edited by G. Pellegrino. Farnham: Ashgate, 133-51.

Kakihara, M. and Sørensen, C. 2002. Mobility. 35th Annual Hawaii International Conference on Systems Sciences 5 HICSS, 131-41.

Kaplan, R. and Kaplan, S. 1989. *The Experience of Nature: A Psychological Perspective*. New York: Cambridge University Press.

Kaplan, R., Kaplan, S., and Ryan, R.L. 1998. *With People in Mind: Design and Management of Everyday Nature*. Washington DC: Island Press.

Kasarda, J.D. and Lindsay, G. 2011. *Aerotropolis: The Way We Will Live Next*. New York: Ferrar, Straus & Giroux.

Kauffman, R. and Techatassanasoontorn, A.A. 2009. Understanding early diffusion of digital wireless phones. *Telecommunications Policy*, 33, 432-50.

Kaufmann, V. 2002. *Re thinking Mobility: Contemporary Sociology*. Aldershot: Ashgate.

Kaufmann, V. 2004. *Motility: A key notion to analyse the social structure of second modernity?* Paper to the mobility and the cosmopolitan perspective workshop, Munich. Reflexive Modernization Research Centre.

Kaufmann, V. 2009. Mobility: Trajectory of a concept in the social sciences, in *Mobility in History: The State of the Art in the History of Transport, Traffic and Mobility*, edited by G. Mom, G. Pirie, and L. Tissot. Neuchâtel: Alphil, 41-60.

Kaufmann, V. 2011. *Rethinking the City: Urban Dynamics and Motility*. Lausanne: EPFL Press.

Kaufmann, V., Bergman, M.M. and Joye, D. 2004. Motility: Mobility as capital. *International Journal of Urban and Regional Research*, 28, 745-56.

Kaufmann, V. and Montulet, B. 2008. Between social and spatial mobilities: The issue of social fluidity, in *Tracing Mobilities: Towards a Cosmopolitan Perspective*, edited by W. Canzler, V. Kaufmann, and S. Kesselring. Aldershot: Ashgate, 37-55.

Kearns, G. and Philo, C. (editors). 1993. *Selling Places*. Oxford: Pergamon Press.

Kellerman, A. 1984. Telecommunications and the geography of metropolitan areas. *Progress in Human Geography*, 8, 222-46.

Kellerman, A. 1989. *Time, Space and Society: Geographical-Societal Perspectives*. Dordrecht: Kluwer.

Kellerman, A. 1990. International telecommunications around the world: A flow analysis. *Telecommunications Policy*, 14, 461-75.

Kellerman, A. 1991. The decycling of time and the reorganization of urban space. *Cultural Dynamics*, 4, 38-54.

Kellerman, A. 1993. *Telecommunications and Geography*. London: Belhaven (Wiley).

Kellerman, A. 1999. Leading nations in the adoption of communications media 1975-1995. *Urban Geography*, 20, 377-89.

Kellerman, A. 2000. Phases in the Rise of Information Society. *Info*, 2, 537-41.

Kellerman, A. 2002. The Internet on Earth: A Geography of Information. London: Wiley.

Kellerman, A. 2006a. *Personal Mobilities*. London and New York: Routledge.

Kellerman, A. 2006b. Broadband Penetration and its Implications: The Case of France. *Netcom*, 20, 237-46.

Kellerman, A. 2007. Cyberspace classification and cognition: Information and communications cyberspaces. *Journal of Urban Technology*, 14, 5-32.

Kellerman, A. 2008. International airports: Passengers in an environment of 'authorities'. *Mobilities*, 3, 161-78.

Kellerman, A. 2009a. End of spatial reorganization?: Urban landscapes of personal mobilities in the information age. *Journal of Urban Technology*, 16, 47-61.

Kellerman, A. 2009b. Global economic crisis, Information society, and personal mobilities. *Environment and Planning A*, 41, 2033-6.

Kellerman, A. 2010a. Business travel and leisure tourism: Comparative trends in a globalizing world, in *International Business Travel in the Global Economy*, edited by J.V. Beaverstock, B. Derudder, J. Faulconbridge, and F. Wiltox. Farnham: Ashgate, 165-75.

Kellerman, A. 2010b. Mobile broadband services and the availability of instant access to cyberspace. *Environment and Planning A*, 42, 2990-3005.

Kellerman, A. 2011. *Mobility* or *mobilities*: Terrestrial, virtual and aerial categories or entities? *Journal of Transport Geography*, 19, 729-37.

Kellerman, A. 2012. Potential mobilities. *Mobilities*, 7, 171-183.

Kellerman, A. and Paradiso, M. 2007. Geographical location in the information age: From destiny to opportunity? *GeoJournal*, 70, 195-211.

Kern, S. 1983. *The Culture of Time and Space 1880-1918*. Cambridge, MA: Harvard University Press.

Kesselring, S. 2006. Pioneering mobilities: New pattern of movement and motility in a mobile world. *Environment and Planning A*, 38, 269 79.

Kesselring, S. 2008. The mobile risk society, in *Tracing Mobilities: Towards a Cosmopolitan Perspective*, edited by W. Canzler, V. Kaufmann, and S. Kesselring. Aldershot: Ashgate, 77-102.

Kesselring, S. and Vogl, G. 2004. Mobility pioneers: Networks, scapes and flows between first and second modernity. Paper tot the mobility and the cosmopolitan perspective workshop, Munich. Reflexive Modernization Research Centre.

Kesselring, S. and Vogl, G. 2008. Networks, scapes and flows: Mobility pioneers between first and second modernity, in *Tracing Mobilities: Towards a Cosmopolitan Perspective*, edited by W. Canzler, V. Kaufmann, and S. Kesselring. Aldershot: Ashgate, 163-80.

Kesselring, S. and Vogl, G. 2010. '…Travelling, where the opponents are': Business travel and the social impacts of the new mobilities regimes, in *International Business Travel in the Global Economy*, edited by J.V. Beaverstock, B. Derudder, J. Faulconbridge, and F. Witlox. Farnham: Ashgate, 145-62.

Kitchin, R. 1998. *Cyberspace: The World in the Wires*. Chichester: Wiley.

Knight, J. 2006. *Higher Education Crossing Borders: A Guide to the Implications of the General Agreement on Trade in Services (GATS) for Cross-border Education*. Vancouver and Paris: Commonwealth of Learning.

Knorr-Cetina, K. and Bruegger, U. 2002. Global microstructures: The virtual societies of financial markets. *American Journal of Sociology*, 107, 905-50.

Knowles, R.D. 2006. Transport shaping space: Differential collapse in time-space. *Journal of Transport Geography*, 14, 407-25.

Knox, P.L. 1995. World cities and the organization of global space, in *Geographies of Global Change: Remapping the World in the Late Twentieth Century*, edited by R.J. Johnston, P.J. Taylor, and M.J. Watts. Oxford: Blackwell, 232-47.

Kopomaa, T. 2000. *The City in Your Pocket: Birth of the Mobile Information Society*. Helsinki: Gaudeamus.

Kshetri, N.B. 2001. Determinants of the locus of global e-commerce. *Electronic Markets*, 11, 251-7.

Kulendran, N. and Wilson, K. 2000. Is there a relationship between international trade and international travel? *Applied Economics*, 32, 1001-9.

Kulendran, N. and Witt, S.F. 2003. Forecasting the demand for international business tourism. *Journal of Travel Research*, 41, 265-71.

Kwan, M-P. 2001. Cyberspatial cognition and individual access to information: The behavioral foundation of cybergeography. *Environment and Planning B*, 28, 21-37.

Kwan, M-P. 2007. Mobile communications, social networks, and urban travel: Hypertext as a new metaphor for conceptualizing spatial interaction. *The Professional Geographer*, 59, 434-46.

Lannoy, P. 2003. L'Automobile comme objet de recherché: Chicago 1915-1940. *Revue Française de Sociologie*, 44, 497-529.

Lapidoth, R. 1994. Autonomy: potential and limitations. *International Journal on Group Rights*, 1, 269-90.

Larsen, J., Axhausen, K.W. and Urry, J. 2006. Geographies of social networks: Meetings, travel and communications. *Mobilities*, 1, 261-83.

Larsen, J., Urry, J. and Axhausen, K. 2008. Coordinating face-to-face meetings in mobile network societies. *Information, Communication and Society*, 11, 640-58.

Larsson, M. and Götaland, R.V. 2009. *Major Trends in Modal Split: Passenger and Freight Transport EU and North Sea Regions.* A progress report of the statistical mapping task of the transport group, NSC. http://www.northseacommisson.eu.

Lash, S. and Urry, J. 1994. *Economies of Signs and Space*. London: SAGE.

Lassen, C. 2006. Aeromobility and work. *Environment and Planning A*, 38, 301-12.

Lassen, C. 2010. Individual rationalities of global business travel, in *International Business Travel in the Global Economy*, edited by J.V. Beaverstock, B. Derudder, J. Faulconbridge, and F. Wiltox. Farnham: Ashgate, 177-94.

Laurier, E. 2004. Doing office work on the motorway. *Theory, Culture and Society*, 21, 261-77.

Lefebvre, H. 1991. *The Production of Space*. Oxford: Blackwell.

Lévy, J. 2000. Les nouveaux espaces de la mobilité, in *Le Territoires de la Mobilité*, edited by M. Bonnett and D. Desjeux. Paris: PUF, 155-70.

Li, F., Whalley, J. and Williams, H. 2001. Between physical and electronic spaces: The implications for organizations in the networked economy. *Environment and Planning A*, 33, 699-716.

Liben, L.S. 1991. Environmental cognition through direct and representational experiences: A life-span perspective, in *Environment, Cognition, and Action: An Integrated Approach*, edited by T. Gärling and G.W. Evans. New York: Oxford University Press, 245-76.

Licoppe, C. 2004. 'Connected' presence: The emergence of a new repertoire for managing social relationships in a changing communication technoscape. *Environment and Planning D: Society and Space*, 22, 135-56.

Limmer, R. 2004. Job mobility and living arrangements. Paper presented at the mobility and the cosmopolitan perspective workshop. Munich Reflexive Modernization Research Centre.

Ling, R. 2004. *The mobile connection: The cell phone's impact on society*. Amsterdam: Morgan Kaufmann.

Lloyd, J. 2002. Departing sovereignty, *Borderlands* 1. http://www.borderlandsejournal.adelaide.edu.au/vol1no2_2002/lloyd_departing.html.

Loewenstein, G. 2002. Psychology of curiosity. *International Encyclopedia of the Social and Behavioral Sciences*. http://www.sciencedirect.com.

Lomasky, L.E. 1997. Autonomy and automobility. *Independent Review*, 2, 5-28.

Lyon, D. 2003. Airports as data filters: Converging surveillance systems after September 11th. *Information, Communication and Ethics in Society*, 1, 13-20.

Macdonald, K. and Gieco, M. 2007. Accessibility, mobility and connectivity: The changing frontiers of everyday routine. *Mobilities*, 2, 1-14.

MacKenzie, D. and Wajcman, J. 1999. Introductory essay: The social shaping of technology, in *The Social Shaping of Technology.* 2nd Edition, edited by D. MacKenzie, and J. Wajcman. Buckingham and Philadelphia: Open University Press, 3-27.

McKenzie, R.D. 1927. Spatial distance and community organization pattern. *Social Forces*, 5, 623-7.

Malecki, E.J. and Moriset, B. 2008. *The Digital Economy: Business Organization, Production Processes, and Regional Developments.* London and New York: Routledge.

McLuhan, M. 1964. *Understanding Media: The Extensions of Man.* New York: Macmillan.

Manderscheid. K. 2007. Book review for Kellerman Aharon (2006), *Personal Mobilities. Swiss Journal of Sociology*, 33, 1164-6.

Martinotti, G. 1994. The new morphology of cities. *MOST* (Management Of Social Transformations) Discussion Paper series 16. http://www.unesco.org/most/martinot.htm.

Massey, D. 1993. Power-geometry and a progressive sense of place, in *Mapping the Futures: Local Cultures, Global Change,* edited by J. Bird, B. Curtis, T. Putnam, G. Robertson, and L. Tickner. London: Routledge, 59-69.

Massey, D. 1994. *Space, Place and Gender.* Cambridge: Polity Press, 1994.

Massey, D. 2005. *For Space.* London: Sage.

Meier, R.L. 1962. *A Communications Theory of Urban Growth.* Cambridge, MA: MIT Press.

Melbin, M. 1987. *Night as Frontier: Colonizing the World after Dark.* New York: The Free Press.

Merriam-Webster Online. 2010. Movability. http://www.merriam-webster.com/dictionary/movability.

Merriman, P. 2004. Driving places: Marc Augé, non-places, and the geographies of England's M1 motorway. *Theory, Culture and Society*, 21, 145-67.

Michener, H.A., DeLamater, J.D., and Myers, D.J. 2004. *Social Psychology* 5th ed. Belmont, CA: Thomson-Wadsworth.

Mijksenaar, P. 2003. Signs of the times. *Airport World*, 8 (4). http://www.mijksenaar.com/pauls_corner/index.html.

Mitchell, W.J. 1995. *City of Bits: Space, Place, and the Infobahn.* Cambridge, MA: MIT Press.

Mitchell, W.J. 2000. *e-topia: Urban Life, Jim – But Not as We Know It.* Cambridge, MA: MIT Press.

Mok, D., Wellman, B. and Carrasco, J. 2010. Does distance matter in the age of the Internet? *Urban Studies*, 47, 2747-83.

Mokhtarian, P.L. 2000. Telecommunications and travel. In Transportation in the new millennium. Washington, D.C.: Transportation Research Board. http://www4.nationalacademies.org/trb/homepage.nsf/web/millenium_papers.

Mokhtarian, P.L. and Chen, C. 2004. TTB or not TTB, that is the question: A review and analysis of the empirical literature on travel time (and money) budgets. *Transportation Research A*, 38, 643-75.

Mol, A.M. and Law, J. 2000. *Situating Technoscience: An Inquiry into Spatialities.* Lancaster: Centre for Science Studies, Lancester University.

Morse, M. 1998. *Virtualities: Television, Media Art, and Cyberculture.* Bloomington: Indiana University Press.

MPI. 2007. http://www.gccoe.mpiweb.org/CMS/mpiweb/mpicontent.aspx?id=10060.

Murakami Wood, D. and Graham, S. 2006. Permeable boundaries in the software-sorted society: Surveillance and the differentiation of mobility, in *Mobile Technologies and the City*, edited by M. Sheller and J. Urry. London: Routledge, 177-91.

NTIA (National Telecommunications and Information Administration). 2002. A nation online: How Americans are expanding their use of the Internet. http://www.ntia.doc.gov/ntiahome/dn/anationonline2.pdf.

Newscentral 24. 2007. http://www.a2mediagroup.com/?c=164&a=17845.

New York, Boston, Washington DC-Media Kit Request. 2007. http://www.vgp.com/advertising/adv_splash.html.

Nick Burcher. 2011. Facebook usage statistics Dec. 31st. http://www.nickburcher.com/2011/01/facebook-usage-statistics-dec-31st-2010.html.

Nielsen. 2007. News Release. http://img2.scoop.co.nz/media/pdfs/0705/carowners.pdf.

Nowotny, H. 2008. *Insatiable Curiosity: Innovation in a Fragile Future*, translated by M.Cohen. Cambridge, MA: MIT Press.

Nunes, M. 2006. *Cyberspaces of Everyday Life.* Minneapolis: University of Minnesota Press.

Paradiso, M. 1999. *Marketing and Place.* Naples: ESI (Italian).

Paradiso, M. 2011. Google and the Internet: A megaproject nesting within another megaproject, in *Engineering Earth: The Impacts of Megaengineering Projects*, edited by S. Brunn. Dordrecht: Springer, Vol. 1, pp. 49-65.

Pascoe, D. 2001. *Airspaces.* London: Reaktion Books.

Pearman, H. 2004. *Airports: A Century of Architecture.* London: Laurence King.

Péruch, P., Gaunet, F., Thinus-Blanc, C and Loomis, J. 2000. Understanding and learning virtual spaces, in *Cognitive Mapping: Past, Present, and Future*, edited by R. Kitchin and S. Freundschuh. London: Routledge, 108-15.

Petroski, H. 2004. Technology and architecture in an age of terrorism, *Technology in Society*, 26, 161-67.

Phillips, R. 2010. The impact agenda and geographies of curiosity. *Transactions of the Institute of British Geographers*, 35, 447-52.

Prato, P. and Trivero, G. 1985. The spectacle of travel. *Australian Journal of Cultural Studies*, 3: 25-43.

Ogden, P. 2000. Mobility, in *The Dictionary of Human Geography*, 4th edition, edited by R.J. Johnston, D. Gregory, G. Pratt, and M. Watts. Oxford: Blackwell, 507.

Olander, L. 1979. Office activities as activity systems, in *Spatial Patterns of Office Growth and Location*, edited by P.W. Daniels. Chichester: Wiley, 159-74.

Oxford English Dictionary 2010. http://dictionary.oed.com/.

Passini, R. 1984. *Wayfinding in Architecture*. New York: Van Nostrand Rheinhold.

Péruch, P., Gaunet, F., Thinus-Blanc, C., and Loomis, J. 2000. Understanding and learning virtual spaces, in *Cognitive Mapping: Past, Present, and Future*, edited by R. Kitchin and S. Freundschuh. London: Routledge, 108-24.

Prendergast, C. 1992. *Paris and the Nineteenth Century*. Cambridge, MA: Blackwell.

Pucher, J. 2004. Public transportation, in *The Geography of Urban Transportation*, 3rd edition, edited by S. Hansen and G. Guiliano. New York: Guilford Press, 199-236.

Raubal, M., Miller, H.J. and Bridwell, S. 2004. User-centred time geography for location-based services. *Geografiska Annaler*, 86B, 245-65.

Raz, J. 1990. Authority and justification, in *Authority*, edited by J. Raz. New York: New York University Press, 1-19.

Redshaw, S. 2008. *In the Company of Cars: Driving as a Social and Cultural Practice*. Aldershot: Ashgate.

Relph, E. 1976. *Place and Placelessness*. London: Pion.

Remy, J. 2000. Métropolisation et diffusion de l'urbain: Les ambiguities de la mobilité, in *Le Territoires de la Mobilité*, edited by M. Bonnett and D. Desjeux. Paris: PUF, 171-88.

Reusser, E., Loukopoulos, P., Stauffacher, M. and Scholz, R.W. 2008. Classifying railway stations for sustainable transitions – balancing node and place functions. *Journal of Transport Geography*, 16, 191-202.

Reuters, 2010. Mother's Day sees highest call volume of year: Study. http://www.reuters.com/article/idUSTRE64611R20100507.

Rheingold, H. 1993. A slice of life in my virtual community, in *Global Networks: Computers and International Communication*, edited by L.M. Harasim. Cambridge, MA: MIT Press, 57-82.

Richards, J. and MacKenzie, J.M. 1986. *The Railway Station: A Social History.* Oxford: Oxford University Press, 1986.

Rifkin, J. 1995. *The End of Work: The Decline of the Global Labor Force and the Dawn of the Post-Market Era.* New York: G.P. Putnam's Sons.

Rifkin, J. 2000. *The Age of Access: How the Shift from Ownership to Access is Transforming Capitalism.* London: Penguin.

RITA (Research and Innovative Technology Administration). 2010. Executive Summary. http://www.bts.gov/cgi-bin/breadcrumbs/PrintVersion_redesign.cgi.

Rodaway, P. 1994. *Sensuous Geographies: Body, Sense and Place*. London: Routledge.

Rodrigue, J.P., Contois, C. and Slack, B. 2006. *The Geography of Transport Systems*. London and New York: Routledge.

Sachs, W. 1992. *For Love of the Automobile*. Berkeley: University of California Press.

Sager, T. 2006. Freedom as mobility: Implications of the distinction between actual and potential travelling. *Mobilities*, 1, 465-88.

Salt, J. 2010. Business travel and Portfolios of mobility within global companies, in *International Business Travel in the Global Economy*, edited by J.V. Beaverstock, B. Derudder, J. Faulconbridge, and F. Wiltox. Farnham: Ashgate, 107-24.

Salter, M.B. 2003. *Rights of Passage: The Passport in International Relations*. Boulder: Lynne Rienner.

Salter, M.B. 2004. Passports, mobility, and security: How smart can the border be? *International Studies Perspectives*, 5, 71-91.

Salter, M.B. 2007. Governmentalities of an airport: Heterotopia and confession. *International Political Sociology*, 1, 49-66.

Scherr Technology. 2009. Internet, mobile, broadband, and social media world usage statistics 2009. http://www.scherrtech.com/wordpress/2009/05/16/internet-mobile-broadband-social-media-usage-statistics-2009/.

Schivelbusch, W. 1986. *The Railway Journey: The Industrialization of Time and Space in the Nineteenth Century*. Berkeley: University of California Press.

Schneewind, J.B. 1998. *The Invention of Autonomy: A History of Modern Moral Philosophy*. Cambridge: Cambridge University Press.

Schrag, Z.M. 1994. Navigating cyberspace – maps and agents: Different uses of computer networks call for different interfaces, in *Telegeography 1994: Global Telecommunications Traffic*, edited by G.C. Staple. Washington, DC: Telegeography, Inc., 44-52.

Schutz, A. and Luckmann, T. 1973. *Structures of the Life-World*. Evanston: Northwest University Press.

Schwanen, T., Dijst, M., and Kwan, M-P. 2008. ICTs and the decoupling of everyday activities, space and time: Introduction. *Tijdschrift voor Economische en Sociale Geografie*, 99, 519-27.

Seamon, D. 1979. *A Geography of the Lifeworld: Movement, Rest and Encounter*. London: Croom Helm.

Seamon, D. 1980. Body-subject, time-space routines, and place-ballets, in *The Human Experience of Space and Place*, edited by A. Buttimer and D. Seamon. London: Croom Helm, 148-65.

Shaw, J. and Hesse, S. 2010. Transport, geography and the 'new' mobilities. *Transactions of the Institute of British Geographers*, 35, 305-12.

Sheller, M. 2004a. Automotive emotions. *Theory, Culture and Society*, 21, 221-42.

Sheller, M. 2004b. Mobile publics: Beyond the network perspective. *Environment and Planning D: Society and Space*, 22, 39-52.

Sheller, M. 2008. Mobility, freedom and public space, in *The Ethics of Mobilities: Rethinking Place, Exclusion, Freedom and Environment*, edited by S. Bergmann and T. Sager. Aldershot: Ashgate, 25-38.

Sheller, M. and Urry, J. 2000. The city and the car. *International Journal of Urban and Regional Research*, 24, 737-57.

Sheller, M. and Urry, J. 2003. Mobile transformation of 'public' and 'private' life. *Theory, Culture and Society*, 24, 107-25.

Shields, R. 1997. Flows as a new paradigm. *Space and Culture*, 1, 1-7.

Shields, R. 2003. *The Virtual*. London and New York: Routledge.

Shin, C.F. and Venkatesh, A. 2004. Beyond adoption: Development and application of a use-diffusion model. *Journal of Marketing*, 68, 59-72.

Shum, S. 1990. Real and virtual spaces: Mapping from spatial cognition to hypertext. *Hypermedia*, 2, 133-58.

Simmel, G. 1969. Sociology of the senses: Visual interaction, in *Introduction to the Science of Sociology* (3rd edition), edited by R.E. Park and E.W. Burgess. Chicago: The University of Chicago Press.

Smerk, G.M. 1992. Public transportation and the city, in *Public Transportation*, 2nd ed., edited by G.E. Gray and L.A. Hoel. Englewood Cliffs, NJ: Prentice Hall, 3-23.

Solnit, R. 2000. *Wanderlust: A History of Walking*. New York: Viking.

Sommer, R. 1969. *Personal Space: The Behavioural Basis of Design*. Englewood Cliffs, NJ: Prentice-Hall.

Sorokin, P. 1927. *Social Mobility*. New York: Harper & Brothers.

Spinney, J. 2006. A place of sense: A kinaesthetic ethnography of cyclists on Mt Ventoux. *Environment and Planning D: Society and Space*, 24, 709-32.

Stalder, F. 2006. *Manuel Castells: The Theory of the Network Society*. Cambridge, UK: Polity.

Statistics Canada. 2004. E-commerce: Household shopping on the Internet. http://www.statcan.ca/Daily/English/040923/d040923a.htm.

Statistics Netherlands. 2004. *Statistical Yearbook of the Netherlands 2004*. http://www.cbs.nl/en/publications/articles/general/statistical-yearbook/a-3-2004.pdf.

Stradling, S.G., Meadows, M.L., and Beatty, S. 2001. Identity and independence: Two dimensions of driver autonomy, in *Behavioral Research in Road Safety*, edited by G. Grayson. Proceedings of the 10th seminar on behavioural research in road safety, 3-5 April, 2000. London: Department of the Environment, Transport and the Regions.

Sydney Media. 2007. http://www.sydneymedia.com.au/html/2285-city-visitors.asp.

Tani, M. 2005. On the motivations of business travel: Evidence from an Australian survey. *Asian and Pacific Migration Journal*, 14, 419-40.

Taylor, J.S. 2005. Introduction, in *Personal Autonomy: New Essays on Personal Autonomy and its Role in Contemporary Moral Philosophy*, edited by J.S. Taylor. Cambridge: Cambridge University Press, 1-29.

Thrift, N. 1996. *Spatial Formations*. London: Sage.

Thrift, N. 2004a. Driving in the City. *Theory, Culture and Society*, 21, 41-59.

Thrift, N. 2004b. Movement-space: The changing domain of thinking resulting from the development of new kinds of spatial awareness. *Economy and Society*, 33: 582-604.

Thorngren, B. 1970. How do contact systems affect regional development? *Environment and Planning A*, 2, 409-77.

Tillema, T., Dijst, M., and Schwanen, T. 2010. Decisions concerning communication modes and the influence of travel time: A situational approach. *Environment and Planning A*, 42, 2058-77.

Toffler, A. 1980. *The Third Wave*. New York: Bantam Books.

Törnqvist, G. 1970. *Contact Systems and Regional Development*. Lund Studies in Geography, Series B, Human Geography. Lund: Royal University of Lund, Department of Geography.

Townsend, A. 2001. Mobile communications in the twenty-first century city, in *Wireless World: Social and Interactional Aspects of the Mobile Age*, edited by B. Brown, N. Green, and R. Harper. London: Springer, 62-77.

Trip, J.J. 2008. Urban quality in high-speed train station area redevelopment: The cases of Amsterdam Zuidas and Roterdam Centraal. *Planning, Practice & Research*, 23, 383-401.

Tuan, Y.F. 1977. *Space and Place: The Perspective of Experience*. Minneapolis: The University of Minnesota Press.

UK Office for National Statistics. 2010. Driving force. http://www.statistics.gov.uk/cci/nugget.asp?id=24.

UNWTO (UN World Tourism Organization). 2006. Arrivals by mode of transport. http://unwto.org/facts/menu.html.

UNWTO (United Nations World Tourism Organization). 2007a. *Compendium of Tourism Statistics: Data 2001-2005*. Madrid, UNWTO.

UNWTO (United Nations World Tourism Organization). 2007b. *Facts and figures*. http://www.world-tourism.org/facts/eng/inbound.htm.

US BTS (Bureau of Transportation Statistics). 2006. *America on the go*. http://www.bts.gov/publications/america_on_the_go/long_distance_transportation_patterns.

US Bureau of the Census. 2006a. *2006 Statistical Abstract: The National Data Book*. http://www.census.gov/compendia/statab/education/higher_education_institutions_and_enrollment/.

US Bureau of the Census. 2006b. *Survey of Income and Program Participation (SIPP)*. http://www.census.gov/population/www/socdemo/workathome.html.

US Bureau of the Census. 2010a. *American FactFinder*. S0804 Means of transportation to work by selected characteristics for workplace 2006-2008. http://factfinder.census.gov.

US Bureau of the Census. 2010b. *2010 Statistical Abstract: The National Data Book*. http://www.census.gov./compendia/statab/2010/tables/10s0593.pdf.

US Bureau of the Census 2011. *2011 Statistical Abstract: The National Data Book*. Table 1054. http://www.census.gov/compendia/statab/.

US RITA. (Research and Innovative Technology Administration). 2010. Table 1-38: Principal means of transportation to work. http://www.bts.gov/cgi-bin/breadcrumbs/.

Uriely, N. 2001. 'Travelling workers' and 'working tourists': Variations across the interaction between work and tourism. *International Journal of Tourism Research*, 3, 1-8.

Urry, J. 1999. Automobility, car culture and weightless travel (draft). Department of Sociology, Lancaster University. Available at: http://www.comp.lancaster.ac.uk/sociology/soc008ju.html.

Urry, J. 2000. *Sociology beyond Societies: Mobilities for the Twenty-first Century*. London: Routledge.

Urry, J. 2002. Mobility and proximity. *Sociology*, 36, 255-74.

Urry, J. 2003a. *Global Complexity*. Cambridge: Polity.

Urry, J. 2003b. Social networks, travel and talk. *British Journal of Sociology*, 54, 155-75.

Urry, J. 2004a. Connections. *Environment and Planning D: Society and Space*, 22, 27-37.

Urry, J. 2004b. The 'system' of automobility. *Theory, Culture and Society* 21: 25-39.

Urry, J. 2007. *Mobilities*. Cambridge: Polity Press.

Verizon Wireless. 2010. http://phones.verizonwireless.com/3g/.

Verstraete, G. and Cresswell, T. (editors). 1999. *Mobilizing Place, Placing Mobility: The Politics of Representation in a Globalized World*. Amsterdam: Rodopi.

Vilhelmson, B. and Thulin, E. 2001. Is regular work at fixed places fading away? The development of ICT-based and travel-based modes of work in Sweden. *Environment and Planning A*, 33, 1015-29.

Virilio, P. 1977. *Vitesse et Politique*. Paris: Galilee.

Virilio, P. 1983. *Pure War*. New York: Semiotext(e).

Virilio, P. 1992. *Rasender Stillstand*. Vienna: Hauser.

Virilio, P. 1998. *Open Sky*. London: Verso.

Visser, E-j and Lanzendorf, M. 2004. Mobility and accessibility effects of B2C e-commerce: A literature review. *Tijdschrift voor Economische en Socilae Geografie*, 95, 189-205.

Wajcman, J. 1991. *Feminism Confronts Technology*. Cambridge: Polity.

Warf, B. 2011. Geographies of global Internet censorship. *GeoJournal*, 76, 1-23.

Watts, L. and Urry, J. 2008. Moving methods, travelling times. *Environment and Planning D: Society and Space*, 26, 860-74.

Webster, F. 2006. *Theories of the Information Society*. 3rd ed. London: Routledge.

Wellman, B. 2001a. Physical place and cyberplace: The rise of personalized networking. *International Journal of Urban and Regional Research*, 25, 227-52.

Wellman, B. 2001b. Physical place and cyberplace: The rise of personalized networking. *International Journal of Urban and Regional Research*, 25, 227-52.

Welz, C. and Wolf, F. 2010. Incidence of Telework. Eironline: Telework in the European Union. http://www.eurofound.europa.eu/eiro/studies/tn0910050s/tn0910050s_3.htm.

Werlen, B. 2005. Cultures in a globalized world. Presentation at the Cultures, Civilization, and Human Development Workshop, Rome.

Wertheim, M. 1999. *The Pearly Gates of Cyberspace: A History of Space from Dante to the Internet*. New York: W.W. Norton.

Wickham, J. and Vecchi, A. 2010. Hierarchies in the air: Varieties of business air travel, in *International Business Travel in the Global Economy*, edited by J.V. Beaverstock, B. Derudder, J. Faulconbridge, and F. Wiltox. Farnham: Ashgate, 125-43.

Wilson, E.O. 1984. *Biophilia*. Cambridge, MA: Harvard University Press.

Wolf, P.J. 2001. Authority: Delegation. *International Encyclopedia of the Social and Behavioral Sciences*, 972-8. http://www.sciencedirect.com/science/referenceworks/9780080430768.

Wolff, J. 1993. On the road again: Metaphors of travel in cultural criticism. *Cultural Studies*, 7, 224-39.

Wood, A. 2003. A rethoric of ubiquity: Terminal space as omnitopia. *Communication Theory*, 13, 324-44.

The World Bank. 2010. Data/Indictors. http://data.worldbank.org/indicator.

Wylie, J. 2002. An essay on ascending Glastonbury Tor. *Geoforum*, 33, 441-54/.

Young, I.M. 1997. *Intersecting Voices: Dilemmas of Gender, Political Philosophy, and Policy*. Princeton, NJ: Princeton University Press.

Young, R. 1986. *Personal Autonomy: Beyond Negative and Positive Liberty*. London: Croom Helm.

Zook, M.A. 2005. *The geography of the Internet industry: Venture capital, dot-coms, and local knowledge*. Malden, MA: Blackwell.

Zook, M., Dodge, M., Aoyama, Y., and Townsend, A. 2004. New digital geographies: Information, communication and place, in *Geography and Technology*, edited by S.D. Brunn, S.L. Cutter, and J.W. Harrington. Dordrecht: Kluwer, 155-76.

Zook, M.A. and Graham, M. 2007a. Mapping DigiPlace: Geocoded Internet data and the representation of place. *Environment and Planning B*, 34, 466-82.

Zook, M.A. and Graham, M. 2007b. The creative reconstruction of the Internet: Google and the privatization of cyberspace and DigiPlace. *Geoforum*, 38, 1322-43.

Zook, M.A. and Graham, M. 2007c. From cyberspace to DigiPlace: Visibility in an age of information and mobility, in *Societies and Cities in the Age of Instant Access*, edited by H.J. Miller. Dordrecht: Springer, 241-54.

Index